writing for **video games**

writing for **video games**

steve ince

A & C BLACK • LONDON

First published 2006

A & C Black Publishers Limited
38 Soho Square
London W1D 3HB
www.acblack.com

ISBN–10: 0–7136–7761–9
ISBN–13: 978–0–7136–7761–4

A CIP catalogue record for this book is available from the British Library.

This book is produced using paper that is made from wood grown in managed, sustainable forests. It is natural, renewable and recyclable. The logging and manufacturing processes conform to the environmental regulations of the country of origin.

Typeset in 10/13pt Bembo

Printed and bound in Great Britain by
Caligraving Ltd, Thetford, Norfolk

CONTENTS

Foreword

Back in the early 1990s, Steve and I first worked together on a game called *Beneath a Steel Sky*. In those days, developing computer games was a rather hit and miss affair. No one really knew how each game would turn out, or even what the design really was. We had a rough idea, and that would do. Our team, at the time, was very small by today's standards – certainly fewer than ten people – and many of those would fulfil several roles on the project. Whoever was deemed to be best at a certain task would get to do it. Steve was originally involved on the art side of things, but quickly expanded his role into puzzle design and writing narrative, as well as helping fend off the criminal elements that regularly found their way into the somewhat seedy office complex we were holed up in at the time.

Looking back, years later, a few things begin stand out that are now worth considering in the context of today's games industry. The first is that team sizes have grown exponentially, along with project budgets. We thought our team was fairly big, but ten people might be a hundred these days. Where that's the case, there's no longer the 'help where you can' mentality. Instead, we see extreme specialisation of roles, and creative writing is part of this, which even has specialisations within it – dialogue, story and plot design, character creation, and so on. What has also changed is that the whole process of designing a game, then implementing it, has become far more efficient and process driven. Schedules have to be extremely accurate, with dire consequences where they are not. The coming wave of Next-Generation games will continue these themes even further as both budgets and risks sky-rocket.

In many ways, when looking at the big titles that dominate the charts in the present day, we can see that some things haven't really advanced so much. In particular, the role of creative writing in the field of computer gaming is hugely under developed. Most games, while being graphical masterpieces that sport ever more sophisticated rendering and physics, are laughably bad when it comes to doing the things both TV and Hollywood have been doing for decades – namely telling good stories with believable characters. To put it bluntly; games are horribly clichéd! Even kids' cartoons are more sophisticated and believable than the macho characters that appear in most computer games. When were you last moved to tears by the death of your favourite game characters, or overjoyed at the plot twist that brings them back to life against all expectations? Most game characters are just a re-hash of what came before, but with better graphics. Their hair might be realistically blown by a fantastically accurate mathematical wind, but the words that they speak often sound like they were written as an afterthought by the programming team.

How has this happened? Why do games lag so far behind other story based media? It's certainly fair to say that there is a serious skills shortage in the domain of creative writing for computer gaming. Historically, most games didn't require great story telling and so while people were learning to program and honing their skills to the lofty standards we see today, no one was sitting beside them investing the same care and thought into narrative. The requirement for great writing skills in mainstream gaming has come about more recently, and the role simply cannot be fulfilled. The problem is exacerbated by the general commercial decline of narrative-based gaming genres such as adventures. Text-based adventures were once big business, but fell by the wayside, in part due to the relentless rise of graphic technology which displaced them as a mass market entertainment form. This is a great pity, not just for adventure gaming itself, but for all the other genres that now have a great need for interactive writing skills.

At this point, you may be wondering why it's clearly proving so hard to retro-fit decent quality narrative back into game development. In truth, writing for interactive entertainment is not easy! As a writer working on a computer game project, you must fully understand the nature of gaming, and of interactivity. The central protagonist in your great novel will do exactly what you want them to – as the writer, you are god. In a computer game, the player expects to be god, doing what he or she wants to do, and in any order they choose. This is a serious headache and turns everything the novelist knows on its head - not many novels make sense if you shuffle the chapters up and read them in random order. To make it work, the writer, working on a game project, must be at the heart of the design team right from the beginning. Too many game projects fail, in terms of narrative, because they try to bring in a "proper writer" too late in the process. This is not enough, and does not work. The role of the writer within the team needs to be taken far more seriously and, of course, we need writers capable of doing this work.

Although it's now over a decade since the days of *Beneath a Steel Sky*, the game is now a cult classic. A team of highly-talented programmers have reverse engineered the original game and recreated it to run on present day machines – not just Windows, but Apple and LINUX machines too. The game is given away free, and it's estimated that hundreds of thousands of players have downloaded it and are enjoying it again. This longevity is highly unusual in the field of gaming, where games disappear from the shelves almost as quickly as they arrive and are just as soon forgotten. The same is true of Broken Sword, the game that Steve and I worked on next. It's as popular as ever, a decade after it was first released. This phenomenon was wholly unexpected. Certainly people are not playing these games for their technical qualities, which were no better or worse than any other game released at the time. What players love are the game characters, and the stories told about the worlds they inhabit. Because these games had writing and narrative right at their heart, they were somehow more real, and alive. Too many games released today are soulless; as a player you can sense that something is missing from them. This must be addressed.

Steve is one of really very few people working in the game industry who not only understands the black-art of writing for games, but can set it out in such a way that you can learn it too. There is a real opportunity here. Good luck – the future of gaming itself depends on you!

Tony Warriner
Co-founder, Revolution Software

Preface

Writers, like other genuinely creative people, are driven by an inherent need to create. Even during those periods when we're not actually at the keyboard pounding out the words, that compulsion will take over our thoughts, distract us in the middle of conversations, and cause ideas to pop into our heads when we really should be paying more attention to our partners sitting across the dinner table. For us, trying to deny that creativity would be akin to denying the urge to breathe.

In order to satisfy the creative urge to the maximum, we will look for ways to push at the boundaries of our creativity; keeping a watchful eye for stimulating opportunities is a must for those of us who wish to find exciting new ways to practice our art. Video games, because of their constantly developing nature, today offer some of the best chances for us to explore our boundaries. For a writer hungry for new challenges, video games offer excellent opportunities to innovate in a medium that devours ground-breaking ideas like no other.

However, if writers are to use existing experience and skills as part of the creation of a successful game, we must all understand how these skills fit into the development process and how the interactive nature of games makes a big difference to what we create. Just as the best film writers understand the process of making a film, so the best game writers must understand the processes involved in developing a game. Only then will we really make the most of our creativity and prove our true worth to the game project.

But how can you, as potential games writers, discover the information you need to find your way into this dynamic field? A search on the internet will be hard pushed to turn up enough correct and relevant information on the subject – at least in the sense to which I am referring. There is much confusion over terminology within the games industry itself and for anyone looking in from the outside it can be more than a little alienating. Too often 'writing games' is taken to mean 'programming games' or 'designing games' and for those of you wishing to enter the field and unfamiliar with the process this can do nothing but add to any misunderstandings you may already have.

Traditionally, much of the writing in games has been done within the development studio's team by the designers or the game's director. In recent years, however, the industry has started the shift towards using specialised writers, who are often brought in from other disciplines – screenwriters and novelists, for example. But what are you to do when faced with the exciting prospect of writing for an interactive medium if you have no idea how the process works? For both experienced writers and those fresh to the field, the use of your traditional writing skills must be placed into the proper context.

Unlike the wealth of screenwriting books that are available, there is a shortage of books that deal with the subject of writing for video games. This book will address that shortage, not only by looking at what makes the subject unique in many ways, but also by putting writing firmly in the context of the game development process and giving you a clear picture of what it is to be a game writer. I will also discuss how the writer's skills are to be adjusted when working in an interactive medium. Many facets of traditional storytelling – plot, character development, conflict, etc. – transfer over to the new media, but need to be looked at with fresh eyes.

Interactivity offers writers, working with development teams, the opportunity to experiment in ways that are impossible or impractical in other media. The information contained here will help you improve the chances of becoming a major part of that exciting frontier and allow you to see you role in the grander scheme of things.

Although this book is aimed at writers who have experience in other fields and wish to develop their skills in a new way and take advantage of the potential opportunities that await them in game development, there is much to be gained for the novice writer, too. In particular, if the book is part of a larger study of writing as a whole, the aspects of writing for an interactive medium covered here will complement other, more detailed writings on such subjects as character, story and conflict.

There has been an increase in the range of university and college courses that cover game development in recent years. Many of the students taking these courses will benefit greatly from the awareness this book can give of a field that's only now beginning to grow.

Writing for Video Games not only has an immediate creative benefit, but for the producers and project managers who must plan development schedules in fine detail, this book will help them see writing as the important set of tasks it is and how vital it can be to weave it into the development of a game in the correct way.

Writing for games is incredibly exciting and rewarding, but it's something that must be fully understood if the maximum quality is to be achieved.

Acknowledgements

No one's skills, abilities and career develop in a vacuum. I certainly wouldn't be in the fortunate position I am without all the highly talented and creative people with whom I've had the great opportunity of working. My heartfelt thanks go out to:

Jenny Ridout – for giving me the opportunity to write this book.

Charles Cecil, Tony Warriner, David Sykes and Noirin Carmody – for the opportunities they game me while I worked at Revolution Software and for everything I learned about game development and design during that time.

Laura MacDonald and Martin Gantefõhr – for ongoing friendship and support.

Neil Richards and Dave Cummins – for their inspirational writing and excellent humour.

Steve Oades – for teaching me some of his wizardry with pixels, palettes and 2D animation.

Eoghan Cahill and Neal Breen – for 2D background art beyond compare.

Francesco Iorio, Jake Turner, Chris Jordan, Andrew Boskett, Patrick Skelton, James Long, Paul Porter and a host of other programmers – for technical brilliance and for putting up with a huge number of design requests and changes.

Ross Hartshorn, Darrell Timms, Jonathan Howard, Dale Strachan and Ben McCullough – for implementation work above and beyond the call of duty.

Mike Ryan, Sucha Singh, Jason Haddington, Mark Thackeray, Steven Gallagher, Alan Bednar, Richard Bluff, Andrew Proctor, Andi Forster, Paul Humphreys, Adam Tween, Jane Stroud, Richard Gray, Linda Smith and others too numerous to mention – for astounding art and animation.

Jan Nedoma, Ard Bonewald, André Van Rooijen, Simon Woodroffe, Dirk Maggs, Renata Richardson, Mike Adams, Bjorn Larsson, Jon Purdy, Burak Barmanbek, Pablo Martin, Chris Bateman, Rhianna Pratchett, James Swallow, Marek Bronstring, Jack Allin, Randy Sluganski, Josh Winiberg, Mathew Meng, Owain Bennallack, Melanie Deriberolles, Mike Merren, Dan Marchant – for opportunities, friendship, feedback and support.

And to all the developers who have created the fantastic games I've enjoyed.

1 Overview

The writer and game development

I was recently approached by a game development studio that wished to use my services as a writer, but had not been their first choice. Encouraged by their publisher, they had initially looked for writers who were established in other fields and contacted a number of them with a view to hiring their services. Unfortunately, because they had no experience of working on games, dealing with those writers was a struggle and certainly didn't work out as the studio hoped. When I was brought onto the project, they readily admitted they were relieved to be working with a writer who understood the game development process.

The problems that had arisen with those other writers had nothing to do with the quality of their writing skills and abilities, but they lacked the specific game development knowledge they needed. The development studio did not have the time or resources to act as nursemaid while the writers learned the ropes and adjusted their skills to fit the new medium and so the working relationship faltered before it even had a chance to begin.

Like other industries, the companies that make up the games industry are governed by the need to create a successful product in order to be profitable and remain in business. However, this is becoming progressively more difficult – as technical and hardware developments become increasingly sophisticated, profit margins are being squeezed to the point where a rigid schedule and budget dictate much of a studio's development process. Anything that is likely to upset that process, add to the schedule or increase the budget will not be considered. This is why it is vitally important for the games writer to not only be familiar with games, but also with the game development process. Only then will the writer be able to bring experience and skills to bear in a way that will benefit the project in an exciting and original manner.

This chapter will look at the industry in broad terms and how the writer can fit into that process. Without this broad view some of the subjects covered in later chapters will not have the right background context.

Small beginnings

Although there were games before it, *Pong* was, when it came out in 1973, arguably the first video game to really capture the public eye. Although it was initially released on machines that were only available in arcades, soon there were versions available for people to play on their own television sets and the home video game industry was born.

Space Invaders and *Asteroids* followed a few years later, once more starting in the arcades, but again soon making the transition to the home. With the introduction of high-score tables with these games, players were now presented with a clear objective – to get on the leader board – something which often fuelled an almost addictive obsession with these games.

A change of emphasis came in 1980 when Activision was formed as the world's first third-party developer and gave their games' individual developers the credit they deserved by printing their names on the packaging. This paved the way for much of the industry as we see it today.

In the same year *Pac-Man* was released and became the first video game with cross-gender appeal. Suddenly women were also playing video games, but this market was something the industry struggled to expand and fully realise the potential it offered. Even today we have an industry that is primarily dominated by tastes of the male player, though some degree of balance has taken place.

The 1980s saw an expansion of gaming through the release of a number of game consoles and the introduction of affordable home computers. The latter introduced a new concept, that of individuals creating their own games from their homes – suddenly anyone who took the time to learn the coding skills had the opportunity to be a gaming entrepreneur. Many of those original bedroom coders went on to greater things within the industry and are regarded by their peers with great respect.

As the 1980s moved into the 1990s, home computers became more commonplace and the quality of gaming improved substantially. Increasingly sophisticated stories could be told and the quality of the graphics improved fantastically. Then, in 1995, Sony released the Playstation and the game market has never been the same again – here was a home video game console that offered superb-quality graphics and gameplay without the need to own a more expensive home computer. Suddenly people who had not previously played games were being drawn into doing so for the first time. Games began to break into the wider public awareness in a way that established them as an entertainment medium to be taken seriously.

The writer's role as a specialist is a relatively recent occurrence in gaming history. Though writing and storytelling appears in games from the early 1980s, it was actually done by the programmer or the designer who put the game together and unless this person had natural writing skills such games were unlikely to be known for the quality of their writing.

The games that formed the initial wave that used the written word extensively were known as 'text-adventures', the first of which was called *Adventure.* Instead of displaying graphics on screen, the player was presented with a series of text descriptions of the locations and played the game by typing instructions – 'Get key' or 'Go north', for example.

The story-based game had arrived and though we did not realise it at the time, the idea that games would employ the skills of a specialist professional writer was being established. The text-adventure gave way to the graphical adventure – combining text and graphics, then later adding animation – and the adventure genre in the early to mid 1990s was one of the most popular at the time. The likes of LucasArts and Sierra created a regular stream of hugely popular titles which were often seen as the cutting edge of computer graphics at the time. It is ironic that today the adventure genre is mostly a niche market that struggles to compete in the larger marketplace which is increasingly driven by the perceived need for costly games.

The legacy of *Adventure* which was furthered by the other developers within the adventure genre has been passed on to the world of video games as a whole. With the advent of cross-fertilisation of ideas and the blurring of 'traditional' game genre boundaries, many games use ideas and styles pulled from a number of sources. It could be argued that the role-playing game has become the main torch-bearer for the story-based game, in terms of popularity, but in today's development climate even high-energy action games are using strong stories and rich characters – feeding the increasingly sophisticated needs of an expanding demographic.

Expansion

The global expansion of the games industry appears to continue unabated. Although there has been some slow down in certain geographical locations, new and developing territories like China, Russia, Brazil, India and others means that the potential growth is enormous and will continue for many years to come. The opportunities for the writer to become a substantial part of that expansion and growth are on the rise, too.

The sales of interactive entertainment software, taken across the globe, reached a staggering £12 billion ($20 billion) in 2004, and have outstripped Hollywood's box office receipts for a number of years. It has been forecast that those sales will at least double by 2007.

The games industry, in just a few decades, has risen from the position where games were often created by teenage coders working in their bedrooms to the point of being run by huge, multi-national publishers, which fund the development of games that often use the skills of a hundred creative individuals and can have budgets of millions of pounds. Skill levels and sophistication have grown in response to the developing hardware and the changing market.

Thousands of individual game titles are released each year that vary in style, size, platform and target market. Game players' tastes vary so much that games are almost impossible to aim at a broad demographic (the elusive 'mass market') but must be treated as a large series of niche markets. Gamers themselves vary from the hardcore and highly skilful to the very casual gamers who like to play accessible puzzle games on their lunch-break or as a way of relaxing. Game writers must not only bring their skills and experience to bear, but must understand the niche they are writing for and the type of gamers who make up that part of the market.

Games were once a market dominated by a young audience – originally seen as the domain of children, not adults – but as that audience has grown older, many of them still want to play games and consequently the demographic has broadened immensely. The average age of the game player is around 30 and rising slowly all the time. Even people who were not children in the early years of the industry are getting into games now that a much wider choice is available, and on platforms that they feel comfortable using.

The expansion of the market is not restricted to an increase in geographical territories and the advent of the new generation of home consoles and computers. There are new and developing outlets for games coming along all the time, including web browsers, mobile phones, PDAs and interactive television.

All game development has restrictions: some are budgetry and time constraints, others are dictated by the conventions of the game's niche and yet others are the limitations of the hardware, such as low memory or lack of graphical sophistication (mobile phones, for example). There is no point trying to write an epic story with a huge cast of characters if you only have 64KB of memory to accommodate the whole game, for example; but being aware of the restrictions helps you to plan how to work within them and become part of the industry's expansion.

Developers and publishers

In its simplest terms, developers create the games and publishers publish, manufacture and distribute them to the retailers. The reality is, of course, a little more complicated than that.

Many development studios cannot fund their own development because the costs are too great and increasing continually – budgets of more than £1 million ($2 million) are becoming ever more commonplace. The publisher, knowing that they need a regular stream of games to publish, steps in to fund a game's development if they believe that the concept is something they will be able to sell to market.

What this means for the studio is that the development of the game is paid for, with the costs set against the royalties the studio would expect to earn from the sale of the game to the public. It also means that the publishers, wanting to keep an eye on their investment, are much more involved in the development process, sometimes reducing some of the creative freedom of the studio.

The publisher will work with the development studio to define the schedule, the budget, milestone deliverables and to define the target market the game is aimed at. The publisher has to be sure they can supply a product that has an expected customer base. If the perceived market is unlikely to exist or is expected to be unreceptive to the game, the publisher may ask the developer to make changes to the concept and design or, at the very worst, may cancel the project after an initial pre-production period.

This can mean that writers may find themselves working with both the developer and the publisher, particularly in the early stages when the concept of the game is being thrashed out.

Many publishers also have their own, internal development teams. This allows them much more control over the creative process, particularly when they are dealing with licensed titles, for which they will likely have paid a lot of money. Though there are those who bemoan the demise of the independent studios, the opportunities for professional writers to work on big budget games are increasing. As potentially big money earners publishers need to ensure that all aspects of the title's development are handled as skilfully as possible.

Increasingly, a good story, strong characters and well-written dialogue are becoming integral to the development of a good game. With game reviews picking up on these aspects more frequently, we are at the point where a studio or publisher would be taking a big risk if they did not hire a professional and dedicated game writer.

The writer's role

Though the whole of this book is aimed at helping those writers who want to work on games understand their role in the development process, a brief overview is in order.

Just to be clear, it is worth mentioning what a writer does not do: a writer does not come up with the idea for the game, write the script and send it to the developer. Where a screenwriter will often create a script and send it to film studios – probably through an agent – there is no equivalent in game development. Game studios usually have more than enough ideas of their own and initial concepts are typically the domain of the game designer rather than the writer. That is not to say that a writer and a designer cannot collaborate in order to create a high concept proposal (even a writer-designer on his own, sometimes) and a number of concepts have been sold this way, but this is not common and is usually something that is done from within the studio where the writer is brought in to work with the team.

Like all others in the development team (programmers, artists, animators, designers, etc.), writers have specific experience and skills which they will use to maximise the quality of all aspects of the project. Writers will not create 3D character models or code the physics engine for the game, but they should be aware of these and other aspects of the project and how the team members' various skills combine to bring about the creation of an exciting and vibrant game. 3D artists will model the characters the writer creates; programmers will develop the dialogue engine that puts the writer's words into the mouth of those characters, and designers will work with the writer to create the gameplay that complements the story. The writer will usually work closely with the game designers, because if a game is to have cohesion, the gameplay and the story should match each other as much as possible.

It is worth noting at this point just what is meant by game design. There has been a certain amount of confusion, particularly in some educational establishments, and the term has sometimes been taken to mean the visual design of a game. Game design is the creation and development of the gameplay. This includes the design of the player interface (the control mechanism for the game), the gameplay rules and mechanics, and how the mechanics are put together in varying combinations to give a satisfying gaming experience. If a story is to be an important part of the game, the design will reflect this – story objectives should match gameplay objectives as much as possible.

The writer needs to be aware of the limitations of the game engine, too. There's no point writing a scene which includes ten different characters if the engine can only handle five speaking characters at the most. Even if the writer feels that this scene is of vital importance, if there is only one instance in the game where this is required the additional time taken to adjust the engine to accommodate a single scene is unlikely to be justified in the schedule and budget.

Being aware of potential issues from other areas of development means that when joining a team the writer can ask all the pertinent questions, which will give a clear picture of the scope of the writing task. How many characters are displayed at any one time? Can the player instigate conversations with other characters or are they triggered automatically based on gameplay or positional criteria? Is the dialogue interactive in any way? How is story information given to the player – through dialogue, on-screen text or some other way? How much character acting for story-telling purposes has been allowed for in the animation budget? Do the characters have a range of facial expressions? Will the dialogue be recorded and is there lip-synching that shows this in the best way? Though these questions will give you an idea of the kind of information it is useful to know when becoming involved in a game project, I will be expanding on it throughout the book.

Above everything else, it is important to understand that, because the subject is the development of games, gameplay is of paramount importance. Even if the story and dialogue are the best things since Shakespeare, they will count for nothing if they swamp the gameplay. The players will probably react against the game because the primary reason people buy games is to play.

Of course, there may be times when the writer knows what their role should be, but the game's director or producer is unclear about what the writer is able to bring to the project. By understanding the development process, writers are able to show more clearly how they can work with the team to enhance the game and increase its chances of succeeding in an increasingly competitive market place.

Professionalism

I hope that what I am going to say here is unnecessary, but if one thing my experience has taught me, it is never to make assumptions if there is the slightest chance that the assumption could be wrong. Like any field of writing, game writing should be approached with a completely professional manner at all times.

Some years ago I was producer on a game that was to become a successful title. Among the reasons for its success was the care we lavished on the project and the attention to detail. An animator was brought onto the team who, although he had considerable animation experience, had not previously worked in the games industry. When, some weeks later, the quality of his work was brought into question, his reply was, 'It's only a game.'

Regardless of any perception of the worth of games in the grand scheme of things, we, as an industry and as individuals, are creating products on which we hope the general public will spend their hard-earned cash. To work to anything less than our full professional standards at all times means we are cheating them of the complete experience they were led to believe they were paying for. If you feel the same way, that 'it's only a game', then perhaps you should think about why you are considering writing for games. If you think that it is an easy route to making quick money, then you will be disappointed.

Another aspect of professionalism is to accept criticism and requests for change with good grace. There will be a great deal of both I can assure you. Some criticism will be genuine and some will be down to misunderstandings or lack of proper communication. Requests for changes, though, can be much more substantial and can range from modifying the story because sections of the game have been re-designed or even removed, to something as significant as the main character is no longer male but female, or the investigative dialogue gameplay has been dropped altogether.

Professionalism also means delivering on time. All aspects of game development are very closely woven within the project schedule and any late delivery could have a knock-on effect. Causing delays in an expensive project is unlikely to win you any popularity contests and it will not be forgotten in a hurry.

All other considerations aside, if you wish to establish yourself as a writer of games, it is in your own long-term interests to be as professional as you can at all times.

The independent route

The traditional development model – developers funded by publishers who deliver the finished product to retail outlets – has come in for a lot of criticism in recent years. The expensive nature of game development and the low royalty rates often mean that studios struggle to make any profit. Increasingly, studios are looking for independent sources of funding that frees them of many publisher ties and where the returns could be much higher.

For the product, the final outcome of this is pretty much the same, however. The game is published, manufactured and distributed to retail stores as before. The major difference is that the studio is able to work with more original concepts and retain creative control. For the writer, to work with a studio going down this route is a potentially more exciting prospect, though the working structure will remain fundamentally the same.

There is, though, another aspect to independent development that is increasing in popularity and scope: when completed, games are delivered directly to the customer through online delivery systems. Put together with small, dedicated teams, these games rarely sell in the numbers of their normal retail counterparts, but with much larger royalty rates many developers are able to make a very comfortable living. For the writers who perhaps want more control over what they create, teaming up with a small collective of like-minded people could offer creative opportunities that other routes would not be able to match.

Playing games

For anyone who works in the games industry, playing games on a regular basis is vital. If you do not play games and, more importantly, if you do not enjoy playing games, how will you, as a writer, be able to relate to the game players and apply your craft in a way that gives them that extra level of quality? How will you begin to understand what works and doesn't in a game if you haven't struggled through weak games and become totally immersed in the good ones? How will you ever grasp the game development process if you do not have an understanding of the end result of that process?

Even if you do not have the time to play whole games – and many involve a huge time investment – you should at least download many of the latest demos and play as large a selection as possible. If you expect or hope to be working on console titles you should buy one of the top game consoles available and play those games and demos to understand the differences between the way they are played and the way that PC games are played.

On the face of it, the untrained eye may not instantly see the differences, and visually there may be little difference between versions of the game. However, the very different methods of interfacing with the game (joypad controller vs mouse and keyboard) can often take a slightly different mind-set to handle it. Very regularly on the consoles, for instance, the control of the main character is applied in a screen-relative mode – moving the controller's stick to the left moves the character to the left of the screen. When screen-relative is used in a PC version of the game it rarely works as well and a character-relative mode is generally more preferable, where pressing the left cursor key causes the character to turn to the left.

One of the most important reasons why the game writer should play games is to get an understanding of why games are – or should be – fun to play. What is it that drives you to complete the mission? Why do you feel great satisfaction at destroying all the ships on this level? Why does the unfolding story feel so much better when the things you do in the game have an effect on how the story moves forward?

Not all games will be enjoyable, of course. Sometimes this will be because the game is weak, though it is often possible to learn as much from a poor game as from a good game. However, it could be that you did not like a particular game simply because you do not enjoy that genre.

Very few people like all styles of games – everyone has different tastes. An excellent sports game may not appeal to someone who enjoys strategy games, but that same person may enjoy a sports management game. With thousands of games released each year it would be impossible to even try to play them all, therefore it makes more sense to identify your own tastes and keep abreast of games that match those tastes. This may, in turn, lead to a kind of specialisation in the games for which you want to write.

The games industry is vast and it is still increasing. The opportunities for writers are increasing, too, but it can be a bewildering industry if you do not know how the development process works. With that in mind, let's move on and discover a little more detail about the writer's relationship to game development.

Interactivity

A fundamental difference between games and most other media is interactivity. Books, films and television programmes, for example, are instances where the reader or the viewer plays a very passive part in the unfolding of the story or the imparting of information. While it is true that the reader will turn the page or the viewer may replay a scene on the DVD, this is only an interaction with the means of delivery and not an interaction with the world or the characters contained within it. The advancement of the plot, the revealing of the story and the development of the character are not reliant on the interaction of the consumer. The experience the creator intended is basically the same for everyone.

Games, on the other hand and by their very nature, are highly interactive from the beginning of play. Not only does the gameplay experience depend on the way the player interacts with the game, the progress through the game relies on the skills of the player, which will vary based upon the type of game played and the difficulty setting the player chooses.

The nature of a game's challenge to the player means that no game can be all things to all players. The hardcore challenges in high-action games regularly fail to appeal to those who prefer a more cerebral challenge or to those whose reactions and dexterity prevent them from mastering the key or button combinations required to develop the game's moves. The labelling of games into types or genres is a hotly-debated topic, but one that enables the potential player to judge whether they are likely to enjoy the gameplay experience or not. Someone who is browsing the shelves of the local game shop for something that will give a blood-filled, action-packed experience wants to be able to find what he or she is looking for without ambiguity.

The writer, like the other members of a development team, must be aware of the degree of interactivity and the style which will be employed in the game. The game should, for the most part, meet the expectations of its target market. This does not mean that games cannot introduce new developments into the genre – in fact, the market almost demands that this happens – but the fundamental gameplay must remain faithful to the type of gameplay that defines the genre. Over time, the genre may change as a result of regular, small changes and many genres are very different now to what they were ten years ago.

Of course, opportunities sometimes come along in which the developer has the chance to create and develop something that is totally new – a game like no other. However, care has to be taken that it is not so different that there is no existing market or one cannot be created. Established game styles develop and change over time, but a product that does not have its roots in a conventional genre is a difficult one to judge and publishers may be reluctant to take a risk – game development can be as frustrating as it is exhilarating.

Because of the huge variety of input devices, gaming platforms and styles of game, interactivity can be an incredibly detailed subject, so rather than miring ourselves in all the permutations that can be thrown up, for our purposes I shall take a broader approach and talk about interactivity from a higher viewpoint.

Passive

For the purposes of this book, passive simply means non-interactive. In other words, the viewer/reader/consumer has no input into the way that the story unfolds, characters develop or the information is delivered. If a customer visits the cinema to watch the latest blockbuster release, from the moment they take their seat until the film is finished, they simply sit, passively allowing the work of the director, camera operators, actors, etc., to deliver the experience to them through their eyes and ears.

The use of the word passive in this sense is not intended to give the impression that the viewer of the film or the reader of the book does not invest a lot of themselves in the experience. A good tale in any medium will create a strong feeling of empathy for the characters, excitement when danger lurks and a whole range of other emotions if the creators have done their jobs well. We, as an audience, will find our hearts beating rapidly during a good horror film and jump at all the scary moments. We will cheer (inwardly, perhaps) as the hero overcomes great adversity and insurmountable odds to win through. We will laugh and cry, feel outrage at injustice and marvel at a genuinely clever plot twist. An excellent story with compelling characters will live on in our minds long after we have left the cinema. At its very best we will find our lives changed by the experience we have just encountered.

In the main, the many excellent qualities of these media are due to their very controlled method of delivery. The author of the novel establishes the pacing, chooses the words used to transport us into this created world and defines our experience. The film maker controls the imagery, the length of the scenes and the emotional turmoil we live through while watching.

Through this control, existing and well-established media have developed in ways that make them incredibly rich and dynamic. Hundreds of millions of people will enjoy a beautiful film, whereas even the best-selling games reach only a fraction of that number. This is not because games are a less valid medium in any way, but that the introduction of interactivity creates a complexity that appeals to different people in different ways.

'Games are not films!' has become almost a rallying cry amongst those who are worried that looking to the film industry for parallels will take games in the wrong direction. Certainly, there is a lot to learn from the skills and experience of those who have developed their careers in other media, but game developers know that they must do so in ways that take nothing away from the nature of their games. For instance, elements like the cinematic use of cameras have been tried in games on numerous occasions with mixed results, mostly due to the imposition of such cameras getting in the way of the gameplay and frustrating the player. We cannot lever in things that will not fit, but we can learn from these other media by understanding their principles and adapting them to work with this interactivity.

Whatever a writer is going to bring to the table when getting involved with a game project, it should only ever be seen as worthwhile if it adds to the interactive experience. Anything that moves the game towards any kind of passive element should generally be avoided.

Interactive

At its most simple, an interactive medium could be one where, for example, the viewer of a television reality show calls in to vote for their favourite contestant. By doing so they are affecting the outcome of the show, though only in conjunction with thousands of other voters.

This type of interaction has a little common ground with games, but I am sure that in general people would never think of the above example as a game in its own right. Most, if not all, games need continual input from the player or players which will feed into and affect the current status of the game. Admittedly, some games, like chess, need lots of thinking time between each interaction, but what separates a game from reality show voting is that the player is responsible for their own part in the experience and what they get out of it.

Most video games have some kind of on-screen representation of the player. This can vary quite remarkably from a humanoid-looking character to a simple cursor or pointer that enables the player to move pieces about the game's playing field. In many games the identification of the player with the on-screen representation – the avatar or player character – is an important part of the gaming experience. Few such representations have become as widely known as Lara Croft, the all-action heroine of the *Tomb Raider* games, but her iconic status is as much a testimony to the importance of strong game characters as to the compellingly interactive nature of the games she appears in.

Interactive software can take on many guises. Feeding columns of data into a spreadsheet programme is interactive, but is hardly something that most of us would call a fun experience. And that is at the heart of what should distinguish a game from any other interactive experience – fun! A game is something with which players should enjoy interacting; something that gets their hearts racing, tests their reactions and skills or makes them laugh out loud.

Creating fun, interactive games is about setting challenges for the players and giving them the means to meet those challenges in a satisfying way. Challenges and objectives should be a mixture of short term and long term. In a simple game like *Pong*, the short term, or ongoing, challenge is to keep the ball in play. The long term objective is to beat your opponent. The style of challenges will depend on the type of game, but will often be quite mixed and varied. Very few games will succeed that are based on a single gameplay mechanic or do not vary the nature of the objectives, so power-ups, better weapons, decreasing time limits and more complex game levels are pretty much par for the course. Interactive variety is very important.

Passive elements do exist in many games, but these are viewed very differently depending on the context in which they are used and the type of player involved. One such element is the cut-scene – a scene triggered by conditions in the game in which the player has no input. These are pre-defined sequences that often appear at the end of levels, act as a game's introduction or convey necessary story information at regular points through-out the game. Depending on the nature of the player, these cut-scenes may be seen as a reward for completing the level or as an obstruction on the way to getting to the next bit of action.

The nature of cut-scenes is changing and becoming more integrated with the regular gameplay. Some game styles are moving towards cut-scenes with an interactive nature, which stops them from being cut-scenes in the strictest sense but keeping the term enables the developer to separate them from the majority of the interaction. Interactive cut-scenes are not entirely new. Most adventure games and many role-playing games have skilfully used dialogue interactivity for a number of years. Adapting such interactive dialogue into more action-packed genres is often a challenge, though, because many action game players see it as a distraction and may choose to skip the scene altogether if they have that option.

As a writer it can be very frustrating to learn that the game has a system which allows the player to skip cut-scenes and even interactive dialogue. Why has the development company gone to the expense of hiring a writer if many of the players are not even going to pay any attention to finely-crafted lines of dialogue or keep up with the developing story? Because the player must have as much freedom to choose their own experience as the developers are able to incorporate into the game. The gameplay experience is everything and if the player feels that the cut-scenes are getting in the way of that experience their enjoyment will be reduced. For the players who want a richness and depth to their experience that can only come from good dialogue, a strong story, excellent graphics and so forth, the work of the writer becomes an integral part of delivering that richness. By offering the option to skip the cut-scenes and dialogue, the game can appeal to more than one narrow group of players and increases the likelihood that the game will be a success.

Some game styles make story and characterisation an integral part of the experience they give to the players. The role-playing game and adventure game genres in particular involve gameplay that leads to the unfolding of the story and regularly necessitates the interaction with other characters in the game to conduct conversations, which in turn give story and gameplay information and objectives. However, even though the players of these games are more in tune with dialogue scenes and cut-scenes, they can still be moved to skip through them if they feel that they are not adding to their enjoyment. This is where the skills of the writer can really come into their own. By creating characters and dialogue of high quality and interest, the players will hopefully find themselves in the situation where they have no interest in skipping the scenes because they add something extra special to the gameplay experience.

Another passive element actually often occurs during the gameplay of many third-person point-and-click games. When the character is directed to move by the player clicking on the game screen with the mouse, a wait of a few seconds takes place while the character moves into position and fulfils whatever the player instructs them to do. With an increasing number of games using direct control and giving the player constant input to the game, the point-and-click style is seen as too passive by many.

This passive interaction has been further exaggerated in recent years with the explosion of storage media and the memory available on the various platforms. Back when games came on floppy discs or small cartridges and computer or console memory was low, the number of game locations was strictly controlled to ensure that the number of discs was kept as low as possible or that the game fitted onto the console cartridge. This meant that to give plenty of gameplay each location had a high number of possible interactions for the player. This high interaction density gives the player a very rewarding experience because there are always lots of interaction points in any given location. With the change to CD-Rom discs and larger game cartridges as the game storage media, the number of locations was no longer so critical, and with huge worlds able to be created in 3D software packages the tendency was towards creating much larger game environments with larger or greater numbers of locations. For the player of the point-and-click third person game, this meant an increasing amount of time where they clicked and waited for the character – although the game worlds increased in size, generally speaking, the amount of gameplay and points of interaction did not increase in proportion. This led to a much reduced level of interaction density (the number of things the player can do at any one time or within any location). Even games where the player has direct control of the character can sometimes feel as though all they are doing is running around large locations, which in many ways is not a fully interactive experience.

Clearly, interactivity is so important that perfectly-valid methods of input are seen by some as less dynamic. This in turn leads to an exploration of new ways to use existing input devices.

One excellent development with mouse control was in the PC first person shooter genre, where the movement of the mouse was directly translated into the movement of the character whose eyes the player is effectively looking through. Suddenly, looking and moving around the 3D environments became so much more intuitive, particularly when combined with other aspects of movement controlled through the keyboard.

Intuitive interfaces are very important because they allow the player to start interacting with the game as swiftly as possible and achieving their enjoyable experience without the hard work of learning a complex interface. This intuitive control should also transfer through to the interface which allows the player to interact with dialogue scenes. If the interface is not intuitive during these scenes, then it is far more likely that the player will be looking for a way to skip the scene altogether, which means that the way the player interacts with the dialogue scene is as important as anything the writer creates and probably more so.

Interactive storytelling

There is much to be learned from established media, and development studios may bring in talented people who understand what works in those media and to combine that knowledge with the skills and experience of the game creators. Such a combination will bring exciting new ventures and give increasingly enjoyable experiences for the players. But as we have already discussed, understanding the nature of games is a key element to the success of any such venture.

Storytelling, and everything that goes with it, has been developed in other media with a high level of sophistication and variety over many years and it would be foolish of game studios to ignore this. But they must also bear in mind that this level of sophistication was developed for audiences that receive their entertainment in a passive manner.

So what of interactive storytelling?

How we want to use interactive storytelling will depend on the type of story we want to tell and the type of game in which we are telling it. Is our story going to be linear or non-linear? Is the player able to interact with and affect the story or the plot, or both?

In linear storytelling, at its most basic, the player interacts with a game in some way that reveals the next piece of the story. If the trigger is the successful completion of a level (defeating all the opponents, say), which launches a cut-scene where the story information and development is shown to the player, the game's story is likely to be linear and mostly simplistic. The purpose of the story in this situation could be little more than a way to link the gameplay sections or create a background setting for the various levels, though it is possible to tell a more involving story if enough of these cut-scenes are triggered. The downside of this method can be to give players the impression that they are not really interacting with the story, which is true, but merely triggering a series of 'chapters'.

This method of delivery is rather like a person walking through the rooms of a house and in each room they enter they find the next pages of a story manuscript. The story is not going to change in any way; the reader has simply had to do some work in order to reveal it. Though few of us would like to read our novels this way, in games it is a valid means of portraying the story, but only if it is handled well. The player must have had an enjoyable gameplay experience of working towards that portion of the story.

This kind of linear story-telling works best when the player feels it has been revealed as a direct result of their actions. For instance, if the player fails to rescue the captured scientist they are unable to discover the story information that the character holds. The progress of the story is directly tied into gameplay success.

Strictly speaking, though, this is not a truly interactive story; it would only be so if the story or plot is changed in some way based upon the way the player interacts with the game and the story it is telling. The more open a game is, the more the player is likely to feel that the story is responding to their input.

Gameplay aside, for the moment, the only way for the story to respond to the actions of the player is if there is a choice the player must make that affects the story. For example, it could be that the player, during an interactive dialogue scene, must choose between lying to the police and telling the truth. Whatever the player decides to do alters the flavour of the game by changing the story or plot, which has now branched. This branching could have a subtle affect that does not affect the gameplay and ultimately does not change the story's ending – in which case the player has interacted with the plot – or it could have a major affect where the whole experience is altered depending on the choice. Gameplay and story could be markedly different in one branch than in the other, which in turn could lead to two very different endings.

A game like this is said to have replay value, because the player could replay the game, make the other choice and play through the other very different branch.

When the mixture of story and gameplay leads to multiple branching points, the potential for the story to have an increasing number of variations – with the amount of work involved escalating exponentially – becomes a very scary prospect. The reality becomes one of controlling the branching, but in a way that gives the players the impression that it is they who direct the unfolding story. This is done by creating a number of branching points that appear to open up the gameplay and story, then regularly bringing together those branches through gameplay and plot requirements.

To expand on our example, if the player lies to the police, later in the game there could be the opportunity to meet up with the detective and tell the truth. While the player might not initially want to do this, there could be a requirement that they do so because some information the detective holds is important to the game progressing. The player's truth becomes a gameplay item that is to be traded for the information and the two branches – truth and lies – are brought back together once more. It may be that this scene could take place at any time throughout the game and play out with a different flavour depending on its position in the story, which would then maximise the openness for the player and the feeling that what they do is having a serious effect.

Alternatively, it may be that this scene is an important, pivotal moment, so it is held back until a certain point in the game in order to pull together a number of different branches. The player character and the detective could trade a number of pieces of information, which then allows the player to move into the final act of the game, in both story and gameplay terms.

What the above shows is that the potential variety for interactive storytelling is enormous and is something that will be expanded upon later.

The vision

Game creation will only ever be successful when the development team are working towards a shared vision. The writer must understand the vision as much as any other team member and much more than some. Without an understanding of the interactive nature of games it is impossible for the writer to even come close to that.

The vision is not simply having a picture in your mind of what the final game is aiming for. It is about understanding how the pieces fit together to give the final game and an understanding of the implications of any changes that are made through the development process.

It is not unreasonable for the writer to expect the section leaders to communicate fully with each other and with the design team and the writer at all times, so everyone knows of any changes, development or progress as the project moves along. Equally, the rest of the team should be aware of the progress the writer is making and the development of the story. Communication is the key to sharing the same vision and ensuring that the development of that vision is seen in the same light by all.

Genres: the game types

The types of games that people play vary enormously and it can be difficult for those who do not play games to appreciate the range of styles that games are divided into. Also, there is a plethora of terminology attached to games, some of which has been purloined from other media with their meanings changed. For instance, the use of the term 'genre' in the games industry has a basic similarity to its use in other media, but it is also very different.

Although genre is a word that refers to a game's type, how those types are defined is where the difference to other media lies. Film genres are defined by subject matter, style of story and sometimes the setting. Common film genres include action, adventure, comedy, western, historical, science fiction, crime, drama, horror, musical and war.

Game genres, however, are defined by the style of play, with little thought for the criteria that define a film genre. In the games industry, then, we have genres and sub-genres like First Person Shooters (FPS), Real Time Strategy (RTS), Role Playing Game (RPG) and so forth, which tells you nothing about the subject matter, but everything a player would need to know to be sure that a game contains the type of gameplay they enjoy. Because gameplay is the most important aspect of a game, it is only right that games should be categorised by their most defining feature, even though some people feel it is wrong to use the term genre in this way when discussing game types.

Subject matter or setting can still be very important – a devotee of First Person Shooters may only like those that have a science fiction setting, where others may prefer such games that have a horror theme. Whatever the game, though, the setting and story style must be presented in a way that fits well with the gameplay style and the genre as a whole.

Those of you who have grown accustomed to the use of genre in other media may find the difference a little disconcerting, yet the underlying meaning emphasises that games are not the same as those other media and the writer (or any other creative person) who wishes to work in the field of games must do so with an appreciation of these differences.

To show the variety of game types, this chapter takes a broad view of the more common game genres; a task not without its potential pitfalls as there is no clearly defined and generally accepted list of game genres. A number of the following headings may well be classed as sub-genres by some people, whereas others may feel that the list should include a greater number of categories. However, the aim is not to create a definitive catalogue of genres but to give you an idea of the many different gaming styles.

The intention, also, is not to define the genres as a set of rules, but to give a flavour of each genre. Whatever game projects you may work on, the fine details can only be gained from the development team.

Action

As a specific genre in its own right, action is possibly less clear than many others. A high proportion of games have an action element – requiring fast responses from the player – so the ones that fall within the action genre tend to be games that have not been included in any of the other genre categories. Action is often combined with other genres to create cross-genres, so you get games that are action-adventures or action-puzzlers, for instance.

At one time, most games that involved the player shooting things or the on-screen avatar running and jumping would have been regarded as an action game, but as games have become more sophisticated, an increasing number of game styles have split off into their own genres (shooters and racing games, for example).

Typical games that fall into the action genre are scrolling shooters, platform games, and maze games, where the player is under constant pressure to keep the avatar safe while navigating towards the game objectives. These games may use 2D or 3D graphics, and regularly offer enemies to defeat or destroy, items to collect, obstacles to overcome and traps to avoid. Most of these games offer some kind of scoring system with bonuses, power-ups and end of level objectives.

Due to the nature of these games the opportunities for the writer can be limited, but there are many action games that include a story of some kind that ranges from the very basic linking levels through on-screen text to stories that tie into the gameplay. The amount of work involved is likely to be small, but the variety on offer can be great fun.

Adventure

Although many consider the adventure genre to be misnamed because so many adventures are relatively sedate affairs, it is a name that has stuck. Because of their emphasis on thought over dexterity skills and taking time over fast reactions, a contrast exists in how the word adventure is used in other media, where adventure usually means an action-packed tale with dynamic characters who are regularly in conflict or danger.

The first games in the genre actually involved text descriptions of locations with the player entering commands by typing on the keyboard. Later, when games appeared where graphics sat alongside the text descriptions, the term graphic adventure was coined which in time was contracted to adventure.

Typically, adventure games are a detective story in a broad sense, where the player must unravel a mystery. By exploring the world, looking for clues, interacting with other characters to find out what they know and by solving the puzzles set as obstacles along the way, the story and mystery will be revealed. Many of the puzzles involve collecting items for use during the exploration and often involve the manipulation of some items within an on-screen inventory, perhaps combining them to create new ones.

Though traditionally, adventures were presented as third-person, the success of games like *Myst* and *The Seventh Guest* showed that a first person view could be emulated through the use of a series of images rendered from pre-defined points. Some consider this to be a sub-genre of its own – one which places the emphasis on (often obscure) mechanical puzzles over character interaction. Some games of this type have moved from the pre-rendered slide show towards real time 3D, which gives for a more immersive experience.

Both first person and third person games have become increasingly well-developed, moving to higher quality graphics, advanced dialogue systems, richer stories, but in spite of these advancements adventure games are now seen as a niche market.

With the increasing importance of games consoles, the personal computer (the mainstay of the traditional style of adventure) has been somewhat relegated in importance. One interesting development is that an adventure created for a console is now a very different affair – one that often includes plenty of action. Although the use of the term in this sense has grown naturally from a different direction, it almost means that the adventure genre has two definitions depending on whether the game is for a computer or a console.

Many of these console adventures have much less emphasis on the detective story side of things and rarely involve complex inventory item manipulation. Inventory items are the things the player collects that are stored in an inventory that the player can access, usually by pressing a specific button or key. Inventory items are often 'used on' other characters or background objects as part of solving a puzzle or overcoming an obstacle (use key on door, for example). Some games have an inventory system that allows you to examine objects while they are in the inventory and sometimes to combine them to make new objects or take them apart to use their components in a new way. They do, however, usually have an increased amount of gameplay involving the exploration of the environment and what the player finds there. Quite regularly they also include some elements from the role playing genre, such as character ability improvement or some element of trade.

Although the console adventure games often have a degree of action, an action adventure in the normal sense is more like a traditional adventure with elements of action placed into the mix. The action is likely to be of secondary importance to the investigative gameplay and will certainly not be something that happens continually.

Although many would probably feel that the survival horror game is a separate genre, it has so many similarities to the action adventure that it is probably best to include it here. For the most part, the survival horror story is one which draws the player character into solving a mystery, with its attendant puzzles and character interaction, while trying to survive the attacks of zombies or other monsters.

For a writer, because adventures have a strong basis in story and character, the genre has the potential to offer rich and involving work. If you are given the opportunity to work with the development team from the beginning you could have a considerable influence of the shape and direction of the story and its characters. In the right circumstances, the chances of creating a truly original story are excellent.

Children's games

Although regarded as a separate genre, games aimed specifically at children are usually games that would fit the other genres which have been adapted to appeal to the target audience, though with proper consideration in the process. Although children's games may be similar to their 'adult' counterparts, the developer must take the age difference into account. Instructions must be simplified and very clear, the interface may not be as complex and the style of reward may need to differ to appeal to a young audience's sensibilities. The survival horror game is not likely to make it through the transition, for instance, but there is no reason not to have a fun but spooky game where the player character is chased by wacky ghouls and ghosts.

Most children's games play to the expectations of the market and are often created with much smaller budgets than their adult counterparts. This often means that the opportunities to create rich stories are few and far between. However, when opportunities do present themselves, writing well within a project's limitation can be a fun challenge for the writer who is able to meet it.

Educational

Educational games are generally designed to make the learning process fun, using the principle that if a child is learning how to count in a game that uses fun cartoon characters from a well-known franchise, say, that child is likely to learn more quickly.

Potentially, educational games could cover a wide range of subjects, but in reality will be limited to those subjects which are appropriate to young children, at whom most of these games are targeted.

Many educational games provide the player with a series of fun activities, rather than puzzles and other more standard gameplay. So an educational art game may have crazy sounds associated with each brush action, say, but could also give the player the opportunity to print out their work.

For a writer who wants to work with educational games, it is important that an understanding of the educational needs of the age group are understood along with accepted teaching methods. Some research may be needed to ensure that the games challenge the child in a fun way and allow them to learn from the experience. Working with developers who specialise in this type of game is very important, bringing in your skills in the right way to benefit the players of the games.

Fighting

This genre is probably one of the narrowest, but is a genre in its own right because no other games are structured in quite this way. Typically pitting two opponents against one another through on-screen avatars, the object of the game is to manipulate the controller in a highly-skilful manner and beat your opposite number. Mastery of the controller by pressing the buttons in specific sequences is an important skill to learn and the fast reflexes required are not to everyone's tastes. All of these games give the player the opportunity to fight against a game-controlled opponent as well as other players.

Because these games are simply about the fighting, the opportunities for the writer are very limited. Some of them will have background or introductory text and descriptions of the various combatants, but their very nature reduces the need for story and character development quite considerably.

Massively multiplayer

These are online games where players pay a monthly subscription to share a virtual world with other players and are often referred to by a number of similar acronyms – MMOG or MMO games (massively multiplayer online games), MMP (massively multiplayer), or MMORPG (massively multiplayer online role playing game).

The origins of these games go back to the 1970s when they were called multi-user dungeons (MUDs), but it is only since the mid 1990s that they have expanded beyond the realm of the simple enthusiast into the phenomenon the genre now is.

Though many games offer online content in some form or another, much of this is either additional downloadable content, the opportunity in a first person shooter to fight against other players instead of the game's AI-driven opponents, or the chance to compete against other players in the latest racing game, to give a few examples. Although these games can be played online, there is usually a relatively small upper limit to the number of players who can play together. Also, this type of online gaming offers little beyond shooting or racing.

Where MMO games differ is that the players' avatars inhabit a persistent world in which they can roam relatively freely and the players are able to choose their own direction in the world. Many games offer combat of some sort, but there are numerous other aspects that immerse the players in these parallel worlds which they inhabit as they play – worlds with social structures and rules of how the players should have their avatars behave.

As a single player in a world populated by thousands of other players, making your mark can be difficult. However, many of the games encourage players to band together, with such cooperative play building a real sense of community within the world. For many people, this kind of gaming experience becomes part of their lifestyle and they will clock up many hours of play each week.

Some MMO games are hugely successful, but the genre is not without its problems. The games, because of their high detail and sophistication, take a long time to develop and are extremely expensive to make. Some projects have been cancelled before ever reaching the marketplace and others have struggled to reach the subscription levels required to maintain a healthy profit. Nevertheless, there is a feeling that after the initial boom period of the genre it is now settling down to a level that is more stable.

The potential for a writer is enormous. Working with the design team to create the world and all the necessary background information is a monumental task and often involves a team of writers to ensure that all aspects are covered. Even after a game is released, there is a need to create ongoing content (content created to keep the game world alive) so that the world is able to offer something on a continual basis to those players who maintain the development of their avatars. New objectives and gameplay scenarios must be delivered regularly if the MMO is to be successful.

Puzzle

Although the pure puzzle game is not, generally speaking, so popular in the normal retail market place, there is a huge demand for puzzle games amongst those players who like to download games they can dip into as a brief diversion.

Puzzle games are generally geared towards using your mind, though many have time limits and speed bonuses so that a certain level of skilful dexterity is also required to progress through the game or to get high scores. Some titles are often described as action puzzle games because they involve the manipulation of moving objects on the game screen within a short space of time. The classic *Tetris* is a prime example of this.

Most straightforward puzzle games occupy the player in some kind of colour-, shape- or pattern-matching that gives the player points, which, as they build up, moves them towards the next level in the game. Many of these games can be highly addictive as the players strive to better their last score or get further into the game.

Word and number puzzles are also a part of this genre, which can range from traditional crosswords to versions of word-based board games where points are awarded for making up words from a small selection of letters. Some maze games also fall into this category where the movement through the maze involves solving it like a puzzle – working out how to manipulate features of the maze to clear the path or to build bridges to the next section, for example.

Writing opportunities in the puzzle genre are likely to be limited. In the main, this is because the games require little in the way of written text or dialogue, but it is also because many of these games are created by small teams on low budgets that are unlikely to stretch to the services of a specialist writer.

Racing

The racing genre covers a wide variety of games – if the game is based on two or more avatars racing along a course of some kind as its prime gameplay, then it is a racing game. Racing games generally involve the player controlling an avatar that represents a vehicle of some sort – cars, hovercraft, boats, spaceships, etc. – though human avatars will be used when it is a snow-board game, for instance. The player may race against other players, game-controlled opponents or against the clock in a time trial mode.

Many racing games offer an increasingly authentic experience with photo-realistic graphics, highly developed physics and attempts to match the handling of their real-world counterparts. Alongside racing, additional game-play features often allow the player to buy new vehicles or upgrade their old ones by spending the virtual prize money they win in races.

There are racing games that avoid the real-world feel and are intended to be simply fun or wacky. Some racing games also allow the vehicles to attack one another as part of the tactics of the race.

Although some racing games have career progression as a part of the game, most are unlikely to need the services of a writer. Though there is often a large amount of text covering details of vehicles and upgrade parts, this tends to be highly technical, rather than creative, and is probably already well-covered by the development team.

Role-playing game

This genre is large, with many titles building into a series of games that become major franchises. Visual and gameplay styles vary considerably, but the underlying feature is that the player controls a character or team of characters and develops their abilities and skills as they explore the game world and the story unfolds.

Characters in role-playing games (RPGs) are governed by statistics. Their skills and abilities are defined by numbers, so the player may control a character that has, say, high intelligence but low dexterity among their abilities. Often teams are built in a way which the characters' skills and abilities complement one another.

As the player progresses through the game, the characters are awarded experience points for successfully using their skills, completing quests or defeating enemies in battle. These experience points are then used to enhance skills or abilities when enough points have been earned to do so, perhaps when the character increases to the next experience level.

The way that role-playing games establish the main player character tends to fall into two camps. The first allows the player to define their own character by choosing their appearance, gender, type, abilities and skills. The second has a main character that is already pre-defined and, generally speaking, the game's story is centred on them.

Story and character interaction are important aspects of an RPG and is often quite involved as the player uncovers it piece by piece. Because of the sprawling nature of many RPGs, they normally enable the player to keep track through some kind of quest or objective screen where anything that is current is listed. Often the player is given the opportunity to take on tasks, or side quests, that are not part of the main story, but offer the opportunity for monetary gain or the chance to trade information or items.

Magic is often a major feature of a role-playing game, particularly those with a fantasy setting. Members of the player's team may well have different magical abilities and styles of spells they can cast; part of the gameplay can involve the learning of new spells and how best to manage magic during battle. In a science fiction setting, magic is often replaced by an equivalent, which could be mutant psi-powers or some kind of mystical force that is magic in all but name.

Combat is a very important part of RPGs, but the style of combat and the interface varies a great deal. A number of games have a combat system that is referred to as turn-based, which means that the battles are highly stylised with each character and opponent attacking in turn. Other games employ real-time combat in which the player is directly involved in hacking with the sword, shooting the laser pistol or casting the magic spell at the same time as the opponents are doing all of these things. Yet another combat style is one in which, once the player chooses to engage the opponents, the game engine takes over and plays out the battle based upon the statistics of all the combatants involved. Players can guide the course of these battles by giving instructions to their team of characters, which could be ordering a specific attack or switching to fighting defensively.

All combat is governed by the characters' statistics, which means that as they improve their skills and abilities they will also be more proficient at combat. Generally speaking, to create a suitable challenge for the player, as the characters improve so do the opponents, either with higher stats of their own or through increased numbers.

Because RPGs tend to be large games, with players regularly investing 40 hours or more in playing them, the opportunities for the writer are good. Sprawling worlds, a large cast of characters, an involving story and numerous side quests mean that a team of writers are usually involved, working together and with the design team. With many RPGs having 100,000 words of dialogue or more, it is clear that for a single writer it would be an almost impossible task to undertake along with story development, quest objectives, item descriptions, and so forth.

Shooters

It almost seems that as long as there have been computer games there have been a good proportion of them devoted to shooting things. From games where the player defends the Earth from invading alien hordes to those where the player takes on the role of a criminal and shoots the good guys.

The primary gameplay mechanic is to shoot the opponents while keeping your avatar alive. Because this can prove difficult, many shooters offer a number of lives to work through before deeming that the game is over or they include power-ups that enable the player to restore health, improve shields or give bigger and better weapons that take out the opponents more quickly.

Shooters have a great deal of variety and refinement today, with many offering background stories and even character development as part of the process of giving the player an ever greater experience.

The first person shooter (FPS) is probably the most successful and widely known sub-genre, which came into its own in the early 1990s and has dominated gaming in many ways since. Often building upon technical developments of the hardware and clever simulation of such real-life features as gravity physics, these games are increasingly immersive to those who like to shoot enemies in seemingly authentic environments.

Many FPS games also give degrees of artificial intelligence to the enemies so they can adapt to the player's style of play. Others offer up to ten levels of difficulty so that, with some experimentation the player is able to find one which gives them a considerable challenge but not without the opportunity of winning.

Most FPS games now offer online play where you can play with or against others. Some of these are a simple an extension of the single player experience where, instead of the other combatants being controlled by the game, they are controlled by other players.

Increasingly, developers of FPS games are seeing the importance of weaving a story through the single player experience. Although some instances have been a little hit and miss, there are examples where this is implemented in such a way that players lose nothing from the shooter experience and still follow a rich story. Because of the normally high budgets for FPS games, opportunities to innovate in interactive storytelling are possibly higher here than in other genres.

Although similar to the FPS, the third person shooter has a very different feel and style of play. The camera moves with the character throughout, but instead of looking through their eyes, the camera is normally positioned behind and slightly above. The third person shooter has a longer history of story-related gameplay and some may overlap the action-adventure sub-genre.

Platform and scrolling shooters became popular in the 1980s and early 1990s and were 2D games that used combinations of sprites. Many of them had gameplay that was continuous, fast and furious and demanded high levels of skill on the part of the player. This type of game has seen something of a revival in recent years with the increasing significance of the independent developer and the online delivery method.

Another type of shooting game is the vehicle shooter. This can vary from tanks shooting each other to dog-fights in space. Often the games' levels are portrayed as missions the player must complete and sometimes a loose background story exists to offer flavour and context.

A few years ago a writer would have found it difficult to find work in the shooter genre and for many games this may still be the case. We are seeing, though, an increase in the number of development studios that want to offer a richer game experience to their players as they strive to be successful in an increasingly competitive field. Stories and character conflict set in a world with consistent internal logic can add layers that make the immersive qualities ever more fulfilling. The opportunities for the writer to work with the design team on such games is on the increase and if done well can be incredibly rewarding.

Simulation

This genre contains games that are simulations (sims) of various aspects of the real world with which the player does not normally have a chance to come into contact. Flight sims and business sims are perfect examples of where someone can have a taste of what it is like to fly a jumbo jet or build a large business empire from scratch.

Some of the games go beyond our existing world's reality and offer the chance to fly spacecraft or build whole new worlds. While they may seem a little fantastical to be considered as simulations, the style of gameplay and the consistency of the game's rule set mean that if such a reality existed the game would be an excellent simulation of that.

Like the other genres, there are plenty of sub-genres that can offer very different experiences from one another. Generally speaking, though, the gameplay is of a cerebral nature, rather than demanding high reflexes and dexterity, though many include the need to react quickly to adverse situations that occur in order to avert disaster. Gameplay can be quite open-ended in many instances and often success is not measured by achieving specific goals but how well the player has done on the road to a more general goal. Success may not even be a factor at all, particularly where some flight sims are concerned, for instance, and it is the experience that counts more than anything.

Flight sims are designed with the intention of giving the player as genuine a feeling as possible of what it is like to fly a real aircraft. The range of aircraft is quite extensive; the game screen is a representation of the cockpit with instrument panels that change and respond to the simulated world. Flight sims commonly use real world terrain and representations of existing airports at which to take off and land.

World building sims are those which allow the player to build a world from scratch, based on the premise of the game's rule set. Laying out the world and populating it will mean that it will develop in a certain way and it is up to the player to manage this growth in a successful way. Because the player has a high level of influence on the world and its population, these games can also be known as 'god sims'.

In life development sims the player controls the life development of single individuals. These can be human representations (where the player guides them from birth to death, say), alien creatures or domestic pets, which must be fed and exercised like a real pet. The game often gives the opportunity to teach them tricks or other fun activities.

Business sims offer the opportunity to take on some kind of high-powered role where the player can play the stock market without any real world risk or build a large business empire, which could be based on anything from railroads to theme parks.

Vehicle combat sims are very different from those games which are classed as vehicle shooting games. In the combat sims, the emphasis is on a realistic representation of how the vehicles move and engage one another. These types of games include flight combat, ground vehicle combat (such as tanks) and even naval battle combat, which can be based on vessels from thousands of years of history.

The sports management sim puts the player into the role of the manager of a sports club – football or cricket, say – and the task is generally to guide the club through a season in an attempt to get as high up the table as possible. Buying and selling of players, managing budgets and revenues all come into the mix and the results of matches are based upon the managerial decisions made, rather than any direct interaction on the part of the player.

Because of the open-ended nature of the genre, the opportunities for the writer are variable. While there will be a need in many for a writer to be involved, much of this work is likely to be fairly generic with little or no chance for clear character development or specific stories to be told. The best opportunities are likely to arise if you are a particular devotee of this type of game and are able to bring your skills to bear in a unique way.

Sports

This genre includes a vast array of sports types and can include anything from boxing to baseball, from football to American football, and from swimming to surfing. Looking at all the individual sports would take up a chapter on its own, but some of the titles in this genre are regularly amongst the best-selling each year, such is our fascination for sport in general. There are players who only buy a games console to play sports games and sometimes only to play those in a specific sport like football.

Generally, the games attempt to represent the sports as accurately as possible while at the same time ensuring they are fun to play. The player will take the part of one of the individuals or teams and is directly involved in the action of running, kicking or scoring. In team sports, the avatar who the player controls will switch from one to another so that the play stays with or near the ball, say.

Writers could have the opportunity to become involved with the creation of the background information that many of the games include, but it is probably worth pointing out that a clear understanding and interest in the sport in question is likely to help enormously.

Strategy

The strategy genre is a collection of game types that involve a high degree of thinking and planning, the movement of pieces on the playing field and often require the management of resources in some way. The game can be an abstraction, like chess, where the properties and abilities of the pieces are stylised to a high degree, or it can display a highly-detailed representation of historical battles or campaigns on the screen.

The style of representation can vary and the use of both 2D and 3D is common, with the visuals ranging from realistic-looking figures and vehicles to 2D maps that consist of square or hexagonal grids.

Turn based strategy games are those in which each player has an opportunity to move, deploy resources, etc., and then waits until the other player (or the game itself) has taken a turn. Thoughtful play is the key element of this style of game.

Real time strategy titles are those in which the play continues regardless of the input of the player, who must make moves and react to the opposition's moves as a continual and ongoing series of situations take place on the playing field. At the same time as establishing supply lines to his troops, a player could also find that an assault by the opponent's troops must be dealt with.

In both of the above styles of strategy game, the player effectively takes on the role of a general who guides his game pieces from afar. However, there is also a type of strategy game where the player is part of the action. The action strategy game can be a game where the player is in among the action in some way and directing troops towards clear objectives through the use of specific orders, often directed at individual units.

Writers could have a strong role to play during development, particularly if the strategy game has an ongoing campaign element that is tied together with a loose story. For the games based on historical battles, researching background information and presenting it as an integral part of understanding the game world could be an excellent way to add richness to the game.

Other games

There are many other games that probably do not fit into these genres and I realise that I run the risk of missing out something that you particularly enjoy playing, but I hope I have shown enough detail without detracting from the real purpose of the book.

Though there are many opportunities for writers in the games industry, there are far more games or game styles that have limited or no need for a writer than there are those that do. The scope and scale of the industry, that this chapter is intended to show, should be appreciated by any writer interested in the medium.

All writers have their own tastes and preferences and the best advice I can probably give is to play games, learn what appeals to you and which of the genres look like giving you the best opportunities to use your skills.

Game design and writing

The game development process can be quite varied, depending on the size and scope of the project and the way that the development studio works, so it is difficult to define a template for what a writer's role will be for every game in development, now or in the future. However, if we look from the development studio's viewpoint, we can appreciate that if there is an interest in hiring the services of a writer, the studio will be more inclined towards one who can show an appreciation of the design process and work well with the game's lead designer. In most respects, the contributions of the writer can be classified as design content, because story, dialogue, character profiles, etc., should all be created in a way that add to the design of the gameplay. It is therefore important that the lead designer on a game feels comfortable with the writer's game writing skills. With this in mind, then, it is important to look at how writing and design may come together with the development of a game.

The most involved games often require months or even years of a writer's time, on and off; while others may occupy a couple of days here and there, with the occasional half-day polishing task. Even games that appear to be very similar can have greatly varying writing needs, which are dictated by the game's design and how the lead designer is able to make full use of the writer's skills.

A game's design could be the responsibility of a single individual or it could involve a larger team controlled by a lead designer whose job it is to ensure that the game's design is cohesive and consistent. Often, contributions from other departments in the studio, such as programming and graphics, will be an important factor in determining the direction of the design and each department will have varying degrees of influence. Design meetings, therefore, can also be very different from one another. A writer can find himself sitting among a group of people with very different ideas on what are the most important aspects of the development process, which will colour their input. Often, though, the writer will be spared a lot of the in-house discussion and simply meet with the lead designer or design team to work specifically on relevant tasks.

At what point in the process the writer becomes involved depends on the importance the writing plays in the game's design and how confident the team are in taking the game down the design path. For the game that revolves around a strong story the ideal situation would be for the writer to be brought in as soon as the initial concepts and gameplay style have been agreed; he then becomes an integral part of the design process from that point. The likelihood is that when the first discussions take place there is already a clear idea of the game with a thumbnail outline of initial story ideas along with possible concept art that has been created for the main characters and some of the settings.

It is possible that the writer may be brought in much later as part of a well-planned development process where the studio wants to be sure that they can prove the gameplay ideas through prototyping before developing the story and characters. It may be that the studio has not even considered using a writer, but during development feels that the game is lacking something and needs the skills of a writer to work some kind of magic. Or perhaps the developers only want to use a writer to create the game's dialogue because the story is relatively superficial. Many other reasons – as varied as the games themselves – mean that a writer could be invited to join a project at virtually any point in its development.

Whenever the writer is brought in, because gameplay must be seen as the most important aspect, they must be aware of how central the design team are in the process. Programming and graphics are also vitally important to the success of the game, of course, but the requirements of those two areas should be defined by the design needs of the game. That is not to suggest that the design dictates the programming and graphical tasks, because each of those teams' development skills can offer valuable design possibilities and the good lead designer will always be on the lookout for what the other departments can add to a game's design. A good sound designer, too, may possibly inspire original gameplay mechanics. It is the lead designer's task to pull together all these skills and inputs, including those of the writer, and design a truly original game with the aid of the rest of the design team.

It is vital, then, that the writer respects and understands the game design process.

Creativity

There are some who feel that game design cannot be taught, which, if true, would make it difficult for me to explain exactly what game design is. However, in a similar manner to other creative pursuits, game design can be taught to those who have the necessary creativity combined with a love of games and the desire to learn the subject. I am going to assume that even if you do not want to become a game designer you have enough interest and creativity to appreciate game design creation. Many of you may not be able to paint or draw, say, but I'm sure you all appreciate the skills and the creative process behind the act of painting.

All creativity is driven by the individual having a constant stream of ideas with which to work. Central to game design is the ability to come up with enough ideas that can be adapted into gameplay mechanics, either on the part of the individual or during brainstorming sessions. Contrary to what seems to be general belief that ideas are ten-a-penny, my feeling is that only bad ideas are that cheap. Good ideas are a much more valuable commodity. If they weren't we'd never again see a bland film, read a tedious book or play an uninspired game.

Sometimes, though, even a relatively weak idea can be made much better through proper development. Very few great ideas spring into the mind fully formed, ready to be implemented; even the best ones need love and attention to make the grade.

Developing ideas

To illustrate the importance of developing the ideas – the true game design process, as it were – I am going to imagine I have just come up with the concept of Rocket Boots. Not stunningly original, I know, but part of the design process is exploring all possibilities in the hope of creating something good from humble beginnings. So how is this idea to be incorporated into the 3D action adventure game I am helping to design?

There are times when an idea will get us so fired up from the start we begin to work on it immediately. Quite often, though, ideas need to mature and develop, becoming full-bodied as they work away in the subconscious areas of our minds. Giving your mind time to mull over them allows you to separate the wheat from the chaff and discard any that are undeserving of more attention.

As any writer knows, much of the creative process comes in taking those initial ideas and developing them through a series of 'what if' thought processes. What if the player character is wearing wings? What if the ball bounces randomly? What if the alien creatures can absorb the energy from the laser pistols?

It could be that the game designer, myself in our example, asks these questions alone, but a better solution could be to have a brainstorming session to discuss the merits of the idea or multiple ideas.

There may be a feeling that the Rocket Boots idea is a little clichéd, so it is probably best to allocate a maximum amount of time to discussing it, at the end of which if there are no significant developments the idea will be dropped. It is just as important to drop weak ideas, or ideas that cannot be incorporated into the current design, as it is to have the ideas in the first place. If an idea is going nowhere, move on.

The initial brainstorming could look at how to adapt the concept and variations may be suggested: jet boots, spring boots and anti-gravity boots. The last idea strikes the team as worthy of further exploration.

Worried that having the anti-gravity boots active all the time may prove to be a gameplay problem, discussion continues about how to limit their functionality and make that limitation become part of the gameplay. With this in mind, one suggestion is that the boots should be anti-gravity pulse boots. A short burst of anti-gravity would shoot the player character into the air, but they would then be subject to the pull of gravity and fall back to the ground. It would be down to the player to work out how to use that sudden, huge leap to their advantage. This is a good development, but there is now a concern that the player will simply keep jumping their character continually.

The next suggestion involves a modification to the boots so that they take thirty seconds to recharge and the battery packs for them only have ten charges. Finding the battery packs for the boots becomes an additional layer of gameplay. At this stage, any numbers discussed are simply pulled out of the air and would require full game testing and tweaking before a proper gameplay balance is found.

Although the mechanic is a feasible one, we need to be sure that it fits the overall concept of the game. If it is a Victorian mystery adventure, then the idea of the boots would never have been suggested in the first place, but the fact that we entertained the idea to the degree we did implies that the basic concept fits with the game's premise.

The next stage in the development of any mechanic is to assess the impact it will have on the overall gameplay – the level designs and level building – and the other mechanics like shooting weapons. Then there are the specific details that have to be considered – the application of physics during flight, the damage to the player for missing the building's rooftop, etc. While looking at these aspects, the design team must be its own devil's advocate, because if any possible problems are not discovered and ironed out at this stage, then they are bound to surface later, as bugs, when they will be much more costly to remedy.

If the writer becomes part of that early development it is because the design team feel they need to develop the gameplay with a strong link to the story and vice versa. Again, brainstorming is an important part of that process because it allows all those involved to throw anything into the mix, helps to break the ice and enables both sides to get to know the way each other thinks. The energy created by a good brainstorming session will probably throw up many silly or wacky ideas that are inappropriate for the project, but that is all part of the process and as long as no one becomes precious about these ideas it helps the session along by allowing everyone to look at the problem from different perspectives.

Continuing with the anti-gravity pulse boots, what could the writer suggest that would enable the mechanic to tie into the story? The game could be mission-based: the player character's commanding officer has just received word that the enemy is developing new technology that could give them the edge in future battles. The player character is given the task of infiltrating the research base and grabbing what information and technology she can before sabotaging the base. This means the character must discover the location of the pulse boots – and the tech information – before she is able to grab them, which in turn feeds back into the design of the research base level.

The writer has become part of the design process.

Based on the story work and the broad design work, the various levels of the game will be fleshed out bearing in mind the logical flow of the game, which is the way the gameplay flows through the game, the order in which it is played and the requirements to move the game onwards. For instance, only when the player gets the red key can he move forward into the red zone, but the design must keep track of what the player must do to get the red key. The writer, in turn, should be aware of such important gameplay nodes and ensure that the story does not conflict with this in any way, particularly if there is more than one gameplay route to the red key.

Of course, even with relatively open gameplay, if the story only progresses at the main gameplay nodes the story the writer creates will not need to allow for the route the player may take between each node, though if the nodes can be taken in a varied order this will need to be taken into account. The level of writer involvement is dependent on what the design team feels is best, but however much the design and writing touch one another within the game, the writing must always support the gameplay.

The gameplay mechanics and interface can have a strong bearing on how the writer approaches the work for a game. If there is no plan to incorporate a mechanic for interactive dialogue, any scenes in which the characters speak will simply play out in a straightforward manner as cut-scenes, though there may be the need for some subtle variations if the order of play means that what the player has done at that point is not fixed. The opposite side of that is if an interactive dialogue interface is an active part of the gameplay the writer must create the scenes so they can accommodate the player choosing to work through the scene by talking about various subjects in any order. If care is not taken in how these scenes are constructed the characters can come across as stupid, which undermines all of the other hard work you have put into the writing.

The setting

The setting for the game is very important, particularly when working with an original intellectual property (IP). How do you create a compelling universe for the game?

Much of this will depend on the type of IP you want to create. You cannot apply the same criteria to a comedy cartoon platform game as to a dark first person shooter or a fantasy role-playing game. The idea of a compelling universe has very different meanings. *Sonic the Hedgehog* is an excellent IP, but would you ever think of the Sonic universe as compelling in the way that you would that of Half-Life?

The key to creating a compelling universe is to ensure that it embodies the style of the game and that the characters, locations and gameplay all sit well within it and complement one another. If the story is layered onto the game in a superficial way, the game's universe will be superficial, too. That could be part of the charm, of course, with a slapstick game requiring a superficial setting to help the mood.

Therefore, you need to look hard at the type of game IP you are creating and build the detail accordingly. Consistency of detail is important, too, so do not throw anything into the mix that is out of place. Working with the design team on these matters is important to ensure the consistency.

The working relationship

When working with the design team in general and the lead designer in particular, part of the process should be in defining how the relationship will work, what is to be expected of you and which information and data you will need to be able to fulfil your task. Reasonably free access to the lead designer – through e-mail or instant messaging – is important as you will need them to answer your questions or to discuss your ideas.

At all stages of working with the design team, the writer should keep accurate notes to keep track of the game's progress. As well as creating story and character profile documents and dialogue scripts, where relevant, you should also write up and summarise meeting notes, e-mail communications should be archived and any design or story changes clearly flagged.

Major revisions to documents should be saved as a new document – there is always the chance that you may need to return to an earlier version because the recent changes have proved impractical for some reason. All relevant people should be given updated documents as soon as they are available to ensure that no one is working with incorrect data. Clear communication is vital at all times and any lack of clarity or other issues should be raised to avoid misunderstandings that could lead to serious problems further down the line. Always clarify any points which are open to interpretation. If something is ambiguous there is always the chance it could be misinterpreted.

Good design documents are a pleasure to read as they will take you through the game with a clear mental image of how it is intended to be played. But even the best can make assumptions at times that are not always caught by the designers creating the document. Be sure to raise questions with the design team no matter how small the point. It may be a simple oversight you have picked up on, but it could also be a potentially large logic flaw that has serious repercussions.

The reverse of the above also applies and any documents you create will be read and scrutinised for clarity and to ensure a fit with the game design. You must respond to any questions and requests for clarification and be willing to change and adapt your documents to ensure cohesion with the design.

The types of documents a writer will be expected to create will depend on the project and where the writer is brought in. Some typical documents could include, but are not restricted to, the following: pitch proposals, story overview, full story, character profiles, story and game background, story and game timeline, dialogue scenes, help files, and instruction manual.

Only through establishing which documents the writer is expected to deliver will an estimate for the time taken be created. Not only is this vital for the writer to know what he can then charge the studio, it is also important for the studio's project manager, who will use that information as part of the project's scheduling.

Scheduling and the setting of target milestones is a key aspect of developing a game that will be delivered on time and within budget. The lead designer will work with the project manager to define the tasks and workload for the design and story side of the project, part of which is dependent on your estimates for the work you will be undertaking. The estimates you give must be as accurate as possible and you must deliver when you say you will. Not doing so can lead to serious delays in the whole project if what you are due to deliver is something that others rely upon. For instance, if the concept artist cannot work on the characters because you've failed to deliver the character profiles, her work is pushed back, which in turn affects the character modeller as he was waiting for the concept work.

Prompt delivery of quality work will not only ensure that you keep the design team (and others) happy, it is much more likely to encourage the studio to call on your services again.

2 Writing and the development process

Interactive narrative

Academic and theoretical ideas of completely open-ended story-telling and of narrative driven entirely by artificial intelligence have a valuable role to play in terms of thought exercises and as a source of suggestions on how to make gameplay richer. However, many of them are too resource-expensive to work in a way that maintains a wide appeal – a large enough audience must be achieved to make the investment in the coding and resources worthwhile. This chapter will look at how interactive narrative can fit with the current climate of developing games where, generally speaking, the narrative is less important than the gameplay.

People practically never buy games for the narrative alone. It may sway their choice when faced with buying two similar games, but the potential player is likely to be considering the gameplay qualities above all else. Even the small percentage who buy games for the story is still looking for an enjoyable gameplay experience because without it they are unlikely to travel very far into the game. For this reason – that gameplay is paramount – our discussion of interactive narrative will always relate to the heart of the game, the gameplay, even when not specifically stated.

What is it we actually mean by interactive narrative? In a broad sense it is simply that the experience of the unfolding story responds to the actions of the participant. In terms of games, those actions are the gameplay choices the player makes. At any one time, the way the narrative responds could be character-related, plot-related, story-related, or a combination of these.

A large number of games have no character development at all, in a narrative sense. Yes, the characters may unleash new powers or discover additional weapons, but for many games if the character starts out as a time travelling tortoise out to save the universe, at the end of the game the character is still the same time travelling tortoise, even though the universe has been saved, in much the same way that the James Bond films work. Of course, if the gameplay is fun the player does not necessarily mind this lack of character development because it may not be important to the style of game. If the gameplay is not fun, then players will not care about character development because they are unlikely to finish the game.

Some games – mainly role-playing games – give the illusion of character development that is under the control of the player, but often this simply comes down to assigning your own scores to various abilities or changing the colour of the character's hair or choosing their gender. Then during the game, the awarding of experience points allows the player to increase the character's stats further. However, affecting the character's skills and abilities in this way is not true character development in a narrative sense. There is rarely any indication that the underlying nature of the character is being affected by the player's actions.

One of the downsides of any kind of game interaction is the possible resource overheads – the creation of the code or art resources necessary to allow the actual interaction to take place and deal with the consequences. Most games rely on repeated gameplay mechanics because when they are spread throughout the game the costs associated with each one become more easily justified. Whenever the player interacts with the narrative in some sense or other, the development team must look at creating specific resources to deal with the action and the consequence.

If you want to develop the nature of the character based on the choices the player makes properly, resources will have to be available for subtle variations of cut-scenes, the creation of a clever, engine-driven facial animation system, or alternative versions of all dialogue lines to reflect the changes in the character, which in turn will require appropriate changes in the responses of other characters. It is easy to see that, for a game with a complex narrative, very quickly the resources required will grow enormously if there is no mechanism in the game to hold it in check in some way.

Some games studios have implemented systems in which the main character never speaks out aloud. This may mean that supporting characters launch into monologues with little encouragement or the player chooses from a series of queries when talking to other characters, which results in the other character responding without the player character having spoken the line the player chose. One dissatisfying aspect of this is that the player can feel the character they control is a little dumb and that the drama that derives from fully-realised character conflict is somewhat restricted or lost altogether because you only ever hear one of the voices and there is little interplay between the characters.

Another method of dealing with characters in an interactive way is to develop gameplay mechanics that handle their responses in a generic or systemic way. If the player character is in charge of a squad of army troopers, say, there could be a system in place that responds to the actions of the player.

If the player looks after his squad, their respect could show in a series of carefully-crafted generic lines. If the player regularly gets the squad members killed by his tactics or recklessness, the remaining members of the squad could show their belligerence through a very different set of generic lines. The down side to this is that generic lines become unconvincing when they are repeated too often and if the same voice is used for another character after the original character has died, the suspension of disbelief is severely strained.

Because the majority of games are action orientated to various degrees, there is a tendency towards favouring an interactive plot or story over interactive character development. This is possible if the action is included as part of the developing story, but in this case the character is changed as a result of the story change and not in any independent way.

As mentioned earlier, simply having a story in a game does not mean that the story is interactive. Many great games have excellent stories that are completely non-interactive and very linear. Before you embark on creating your game story masterpiece, be sure that the lead designer knows exactly the level of interactivity they want from the narrative so that you can deliver to those specifications.

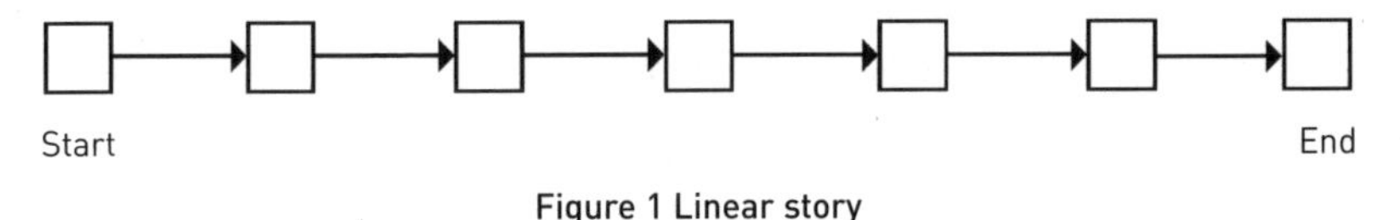

Figure 1 Linear story

In a linear game story (see Figure 1), the progress of the story happens at story nodes, the small squares, with the arrows representing the gameplay that takes the player to that node. Each story fragment is effectively a reward for completing the previous section of gameplay with the number of nodes and fragments being dictated by the depth of the story.

The reward nature of the fragments should not be obvious to the player; they should appear in a logical fashion and any gameplay objectives should match the story objectives. This way the story feels like it is an integrated part of the game design.

Even in gameplay that is completely action orientated you can do this with gentle nudges and reminders of the story and gameplay objective – download a map from a console which shows the layout of the area and the player's objective clearly marked; have the player character's boss call her up on the radio or mobile phone to give her further information or instructions. The possibilities are dependent on the style of game being developed, but by working on how best to maximise the way the story and gameplay fit together, a rich, linear story can be an excellent accompaniment to a strong game. The real beauty of a linear story is that because the studio does not have to allocate time and resources towards creating alternative scenes and dialogue, you can maximise the quality and impact of the scenes you use to tell the tale.

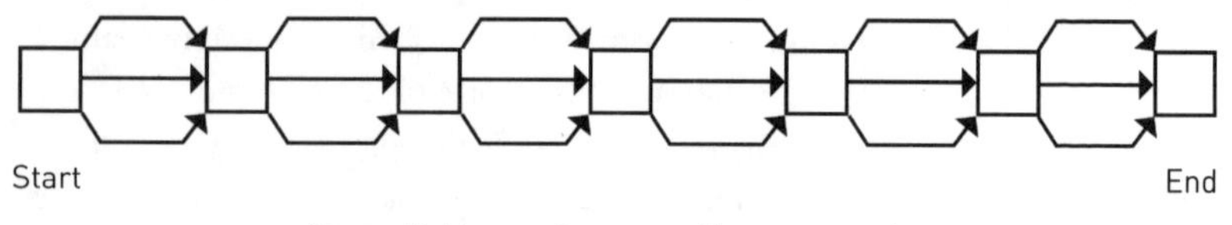

Figure 2 Linear story, non-linear gameplay

Figure 1 not only shows a linear story, but also linear gameplay, meaning that the route the player takes to get through each level of the game is predetermined and fixed. Figure 2 shows a variation where the story is still linear, but the gameplay that leads to each story node is varied and will depend on how the player approaches each level. This usually means that the players have the option to draw from a range of gameplay mechanics in any way they choose, which gives them the feeling that they are in control of the gameplay. This type of game/story structure is a strong combination as it has the advantage of relatively open gameplay with a specific and potentially powerful story.

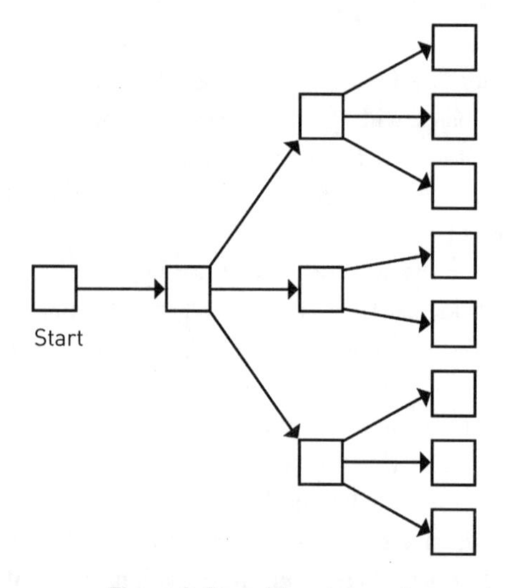

Figure 3 Branching story and gameplay

When you introduce changes to the story and gameplay that are dependent on the choices of the player, the branching lines through the game increase very rapidly. Figure 3 shows that after only a few levels the number of story nodes could be huge, depending on the number of choices the player is offered at each node. Continuing this onto the end of the game could result in a hundred or more possible game endings. In a very pure sense, this would be the ideal situation from a player's point of view – having complete control over the unfolding of the story – but the practicalities make it impossible to create. The resources needed to offer all these possibilities would be beyond the budget of even the largest game. Keeping track of the development of the project and testing all of the possibilities thoroughly would also mean that there would be a lengthy development period.

There clearly needs to be some practical middle ground that allows for more player freedom without that freedom getting out of control.

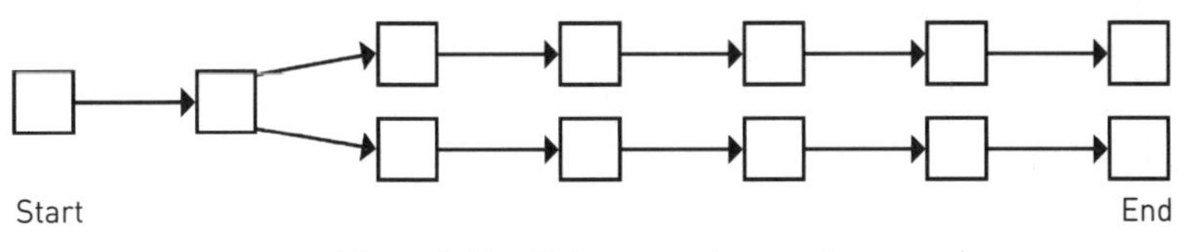

Figure 4 Parallel story and gameplay

Figure 4 shows one possible way to do this – give the player a choice early in the game which leads to two parallel gameplay and story paths. The advantages are that the player is able to choose the main character's destiny but the development team has control over the two strands. This is a very good way for the player to affect the character's development in a significant way without the complexities of an ongoing series of character choices. Just as in Figure 2, the gameplay between each story node can be as open as possible to maximise the player's sense of freedom. The game could also give the opportunity at points along the two strands for the player to make choices to switch to the other path. If this is implemented well it could give the player the impression that they have much more freedom than they actually do.

The disadvantage to this method is that if the player only plays through the game once he or she will only ever see half of what has been created. However, because the game has this alternative gameplay and story, it has potential replay value for players who are interested in playing the game again, but differently. Key to the success of this is that the player finds the game enjoyable enough that they want to play it again. While some players love to replay games, others would rather move onto another game, so replay value may depend on the target audience.

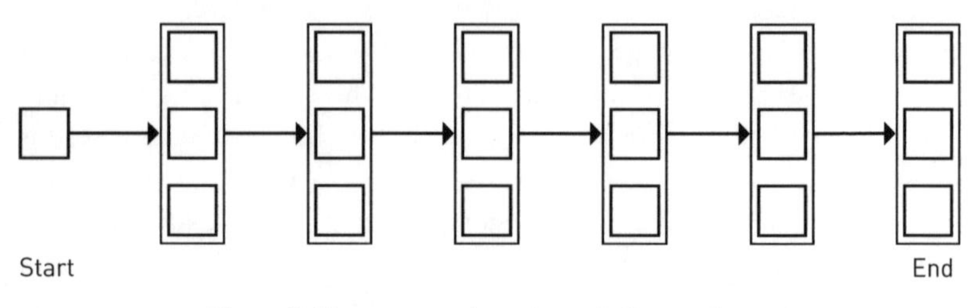

Figure 5 Linear gameplay, player-influenced story

An alternative to the parallel strands is shown in Figure 5. The gameplay levels are largely unaffected by the choices of the player, but at each story node there could be subtle variations that have been affected by the previous story node choices or by the style of play through the last gameplay level. The earlier squad example fits into this structure. This can still be an expensive way to create a story that reacts to the player if the story nodes contain scenes that are pre-rendered full-motion video, but if they are handled within an in-game scene mechanism the overheads can be significantly reduced. Again, the gameplay itself does not have to be as linear as shown in the diagram and can be as open as possible within the confines of the level.

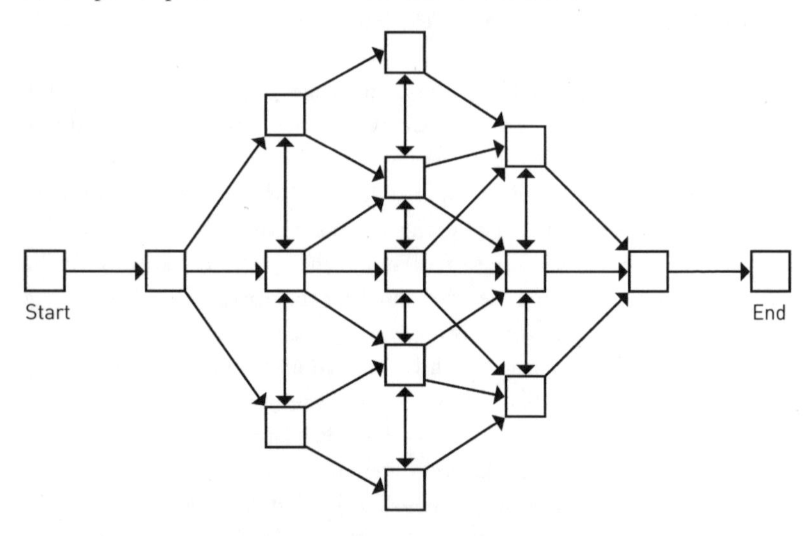

Figure 6 Controlled branching story and gameplay

If the story and gameplay are really to go hand in hand and both become an integral part of what gives the player a rewarding experience, then a controlled branching method may well suit those requirements. Figure 6 shows a possible way that this might be structured.

Although the players are given choices as they make their way through the game, they are very controlled choices which ultimately draw the player back to a single climax. Of course, it is still possible to have multiple endings depending on how tightly you want to draw the threads back together again. If the ending is the same, regardless of the path taken to get there, the story is largely unaffected and it is the plot that is interactive.

The controlled branching model can work in two ways. The first is simply that the player's choices always drive the action forward, so the path towards the ending is defined by the player by tracing along a combination of forward-moving arrows. The player will only ever get to pass through one story node at each level of the game – on the diagram this is seven story nodes in total, but in reality it is likely to be many more.

The second, more complex way that controlled branching can work is to offer the player the choices as before, but also give the opportunity to visit the other choices, too (indicated by the double-ended arrows). In fact, this may be a requirement of the game and although the player could be given the choice of visiting the pub, the gallery or the police station, the player character perhaps has to do all three as part of the game's investigation. In a way it is similar to making pancakes – you need to buy the eggs and milk and flour before you can do so. Of course, the likelihood is that you would not have to go to separate places to buy the individual ingredients, but if you did, then it is up to you whether it is the egg shop, the milk shop or the flour shop you visit first.

Taking the pancake analogy further, it could be that your cooker is broken and needs repairing. Do you get the man in before you buy the ingredients or after? What if there is an option to arrange the repair and then get a friend to wait in for you while you go to the shops? What if the friend calls you up to say that the guy will not do the repair without being paid first?

All these variations and possibilities introduce a level of complexity that involves a lot of work to keep track of, but there is incredible potential for variety and for giving the players a unique experience they have unfolded through their choices and actions. Yet, because you and the design team have created a structure that you control, the richness of the story can be much greater than if there is little control.

One thing that we need to emphasise, here, is that the structures I have outlined are not a definitive list. Not only is it possible to take elements from each and mix them up in any way that you like, but because game development is constantly changing, it means that new structures are likely to emerge as we learn more about the possibilities that interactive storytelling offer us. It could be, for instance, that the game has a central hub cluster that uses a controlled branching structure, but that throws up the need to visit other areas of the game world that are very linear and when completed return the player to the hub.

When looking at traditional storytelling, the act structure of the narrative is often at the forefront of any discussion. Although an act structure can be applied to a game's narrative, the more open it becomes, the harder it is to identify the act boundaries. It could be that a certain key scene may define the end of act one, but if the player has some control over the unfolding plot, other scenes could appear in either act one or act two. The lines become a little blurred and the pacing depends on the direction the player takes.

For games that have a more linear structure, it could be argued that each gameplay level should be treated as an act in its own right because of the climax at the end of each, though it may be tempered if you develop a higher level structure which sits on top with each act containing a number of gameplay levels.

Something I have not seen considered before in relation to story is the idea of player character death. Discussion tends to assume that the end of the story comes with the game's finale, but if the character dies before that point is reached, that is the end of the current story for the player. Starting the game again or re-loading from a suitable save point is effectively the player trying to experience another story with a different ending – one in which the player character wins through to the end of the game. However, because the players consider any death to be a failure, they generally want to push past it and start again, resetting their mind, almost, to cancel out the failure.

The future of interactive narrative is one of the most exciting areas in entertainment. With games and other interactive media spreading across a wide variety of platforms, the need for quality content will increase as more companies deliver interactive content to an expanding number of customer demographics.

Targeting an audience

The gaming audience is gradually growing older with the current average age of the player in the late 20s. It has been reported, though, that the average age of people who buy games is somewhere in the mid 30s. So the prevailing wider perception that games are just for children is, in reality, no longer the case; something which is reflected in the increasing number of 'darker' games that are aimed at a more mature audience. To illustrate further the wider appeal of gaming, it is also said that nearly 20% of game players are over 50 years of age.

The gender appeal is broadening, too, with more than 40% of gamers in the USA being female and markets in other territories beginning to follow suit. There are more female gamers playing online than hardcore gamers, something that is not going unnoticed by many of the online gaming providers. There is still a concern that this broadening of the audience is not being fully embraced by the conventional retail sector and that many high profile, big budget games are still aimed at a young male audience with a preference for adrenalin-stoked gameplay and technology-driven visuals over compelling content. Where are the games that are the equivalent of the chick-lit novels or films, for instance? There is clearly a long way to go before games are created for all market sectors and target audiences with an equal standing.

Although it would be ideal to create a game that reaches this widening demographic, as a writer, you are unlikely to be able to choose the target audience, unless you are fortunate enough to convince a publisher that an untapped market will buy huge numbers of the game based on your ideas. More often than not, the intended market and gameplay style will have been decided when you are brought onto a project, so it is vital that you consult with the development team to ensure that you have a clear understanding of the audience you are writing for.

The mythical mass market

Many games have foundered because they have been created with the idea that they will appeal to a mass market that I do not believe exists. Adding features that the publisher thinks will generate more sales may actually have the reverse effect if those features alienate the core audience.

Interactive experiences are something very personal to an individual, and players quickly decide on the types of games that appeal to them, often buying games from a small selection of genres or sub-genres. However, that combination of styles is likely to be as personal to that player as the games themselves. Two fans of fantasy role-playing games, for instance, are likely to enjoy playing other genres, and while one may prefer adventures and real time strategy games, the other could enjoy first person shooters and car racing games.

Many players will have an interest in other genres outside their core grouping, but only for games that they feel are worth their time. Often, the games that are classed as having a mass market appeal are simply the ones that are so well designed and made they are in the top few percent of their genre. They draw in additional players who would not normally buy that genre's 'run-of-the-mill' games. Publishers who see the success of a genre-defining game may try to copy that success by financing the making of a similar type of game, but only rarely does this actually succeed.

When the installed user base (the people who own a game system) of consoles and computers is considered, there is a potential audience in the hundreds of millions. A game that sells a million copies may be a big success, but it is only reaching a tiny fraction of that audience. For example, we know that the number of Playstation 2 consoles sold is close to 100 million throughout the world. If you add on the other consoles (Xbox, Gamecube, Gameboy, etc.) as well as huge numbers of home computers, we get numbers in the hundreds of millions. If a game sells a million, then it is clearly reaching much less than 1% of the game systems in use throughout the world. It therefore seems natural to consider all games as being part of a series of niche genres. Admittedly, some niches are much larger than others, but if top games ever had true mass market appeal they would sell fifty million copies instead of the five million, which they sell if they are very fortunate. Even the larger genres, like first person shooter games, have a high number of games that sell poorly because, not only do they fail to appeal to those outside of the genre, but they also fail to appeal to many of the players who are fans of the genre and whose expectations are raised by the high quality of the genre's best games.

Occasionally, there are games which appear in the market that seem to have a broader appeal in a way that transcends existing genres, but unless the sales are really huge the game still is not mass market, it has simply reached a large niche audience for which there is currently no defined genre. It could be that the game is the first of a type that defines a new genre within which other games will follow.

Understanding that the audience for your game fits within a niche genre, established or otherwise, will allow you to write in a way that maximises the quality of your game within that niche.

Knowing the audience

Once the genre and the audience for the game have been established, it is important that everyone involved understands the conventions of the genre and the expectations of the audience. If you do not know the target market, how will your writing appeal to them other than by chance?

Playing other games, which are similar in gameplay style to yours, is vital. Both the best and the worst in the genre can teach you valuable lessons in what appeals to the players and what mistakes you can avoid. Familiarisation helps you form a picture in your mind when the design is in the early stages and all you have to go by are the design documents. Understanding how the game is expected to work, based on a few descriptions, is very difficult if you have no frame of reference. Be sure, too, to play the games the designers use as a yardstick in the genre so that you know the kind of game standard at which they are aiming.

Other valuable sources of information are the many reviews, both online and in print, which regularly highlight the problems that games may have. When written well, these reviews can really get to the heart of why a game is successful or not, and how it might have been improved if features had been implemented differently.

The coverage, within reviews, of the writing in games can be a little patchy, with wildly differing opinions on the same game, but this is probably a reflection of the varied quality of writing that has been common to many games. It is difficult for reviewers to develop a consistent approach to reviewing game writing when the overall standard is open to question.

Care must always be taken when reading game reviews. While they are valuable for the insights they can offer into the views of the target audience, at best they only represent a very small percentage of that audience. The person who plays a game to write a review of it is very different to the game player who may spend a couple of hours a week playing, working through a game very slowly. Sometimes, too, a review can conflict with the general feeling of a genre's audience, particularly if a reviewer is not a devotee of that genre – many sites and magazines have a limited number of review staff and games are given to them which may fall outside their normal range of playing styles.

Another potential insight into the target audience can be gained from the online communities that focus on your game's genre. Not only is it valuable to read what the members have to say on the latest releases within the genre, if you have any particular research you wish to undertake many of them are pleased to answer your queries or fill in your questionnaires. However, while their views are valuable, the members of community forums can sometimes be a little too hardcore and become obsessed with fine detail – such is their love of the particular genre – and this often will cloud the bigger picture.

Expanding the audience

Once you know what kind of game the team is creating and you have been able to define your audience and understand them in general terms, is it possible to expand that audience? More to the point, is it possible to appeal to players outside the genre without alienating the core audience within the genre?

This is largely the realm of the design team – how can they add to the gameplay and hold the genre's audience at the same time as giving the game a wider appeal? – although the writing can be an important part of the process and become a key part of such development. For instance, how can you integrate a rich and complex story into a high-action game without detracting from that action? How can character interaction be made to work in game types that traditionally have no such thing? How can narrative become a part of a car racing game?

Unfortunately, there are no simple answers to these and other questions that will arise in the early stages of a game's development. Because each game is unique, even within a genre, the solutions are going to be ones that fit that particular game and which also work within the genre.

Work with the design team to incorporate interface features (elements that often make a game unique – something cool that happens when the player presses a button at a specific time or presses a combination of buttons to create an unusual character move or on-screen effect) that allow you to tell the story in a suitable manner. Define the level of character interaction with the designers and get them to plan the game mechanics in a way which allows you to maximise the quality of the writing within their defined parameters. Look for ways of incorporating the new ideas so that the genre's traditionalists can skip them if they wish and have an experience that matches their expectations of the genre – for instance, when players begin the game it could offer them a choice between a story mode and an action/arcade mode.

Working closely with the design team is an important aspect, but even more so is the need to produce work of the highest quality. As mentioned earlier, only the best games in a genre have a chance of drawing a wider audience from those game players who may not be particularly interested. Anything less will give them the excuse to regard your game as just another average game and pass it by. It might not be possible for you to have any influence over the gameplay design or how the development team will implement all the features, but when it comes to the writing the onus is on you to deliver the goods. Write to enhance the style of the game and in a way that the genre's players will appreciate.

It may be that the intention is not to expand the audience, but to ensure that the game appeals to the maximum number of players within that existing audience. In which case, all of the above still applies and could be a much more realistic target. Instead of the team's focus being diverted towards a wider audience that only exists in potential terms, concentration on making the game as good as possible for a known audience may well give it the quality it needs to draw in a wider audience as a secondary objective.

Getting to know the audience is not only important in understanding how to approach your writing, but is also vital if you are to avoid creating something clichéd or derivative. Only when you know the conventions of the game's genre can you bend and manipulate those 'rules' – even break them at times – to help you create the exciting experience demanded of modern games.

Characters and point of view

Characters have an assortment of different roles in games. They can be a crucial part of the whole gameplay experience or, sometimes (in many puzzle games, for instance), they do not exist at all. Their nature and how they are represented will be determined by the type of game, the style of the interactive world and can range from photo-realistic humans to cartoon hedgehogs to lumbering stylised battle robots. Although it could be said that the vehicles in racing games are a kind of pseudo character, these types of avatars are not part of the discussion in this chapter; here we will assume that a character in a game is defined as an entity that, if the game world were real, would have some kind of mind of its own.

In recent years, the idea of game characters having a mind of their own has become important in certain types of action games. A degree of artificial intelligence has been developed for the player's opponents and also for the team members in squad-based games. Having other characters in the game who either provide an element of unpredictability or have the potential to learn from how the player plays the game, can offer an excellent degree of immersion. Through such means it is possible for the game's balance to match the player's own skills and abilities.

It is important to realise, though, that characters in games are much more than a series of well-coded behaviour traits. In particular, a game's important characters must be treated as you would if they were being developed for a top film or TV series. They should be fully rounded with a relevant back story, clear motivation and a well-defined sense of what makes them tick. You must understand what makes them behave the way they do in the game world.

Background characters

Often, the reason for having characters in a game is governed by a gameplay need – they provide the obstacles the player must overcome. The writer must then work within that constraint to justify their inclusion within the story – the antagonist has ordered his henchmen to ambush the player character because he has been tracking her through the research centre, say, and because he does not trust the competency of the goons he puts his elite force on standby.

It is very easy to think of grunt characters simply as cannon fodder, for in many cases that is exactly what they are, no matter how good their artificial intelligence. To save on resources many of them may look alike, using the same 3D models and textures, which adds to the generic impression given. If a writer can somehow develop a few of them beyond the generic and fill them out a little, the rest of the cannon fodder characters are fleshed out, too, by implication. Overhearing a couple of guards bitching about their boss or discussing football results expands the game's world beyond the immediate and can suggest to the player that these characters have a life beyond their henchmen duties.

Human henchmen are not the only cannon fodder in games, particularly if the game is fantasy, cartoon based or stylised in other ways. In many of these games, an ogre is simply a mindless enemy and a furry blob with no arms has nothing else going for it beyond kill-or-be-killed. Expanding these characters is much more difficult, but if the opportunity to do so is there, it can be an important part of the value a writer brings to the project.

Children's games in particular can be given added value and richness with attention to detail and care in the way even minor background characters are portrayed. One of the important reasons films like *Finding Nemo* and *Toy Story* succeed is the richness of detail in the characterisation. Both children and adults may become enthralled in the fullness of a story that revolves around excellent characters and the same can be achieved with games if they are approached with this in mind.

If a character has a speaking part, even if it is only a small one, you should avoid calling them 'Guard 1' or 'Woman 2' but refer to them as Rick and Mabel, say. If nothing else, giving them a name lets you see them as something more than just a non-entity and also enables the actors at recording time to put a little more into the few lines the character has. Choose a name, then, that will work with the lines and give a hook the actor can work with.

Interactive characters

Stepping up from the cannon fodder we get to the characters who are a little more special. They are one-off characters, level bosses (extra powerful opponents who often appear at the end of a gameplay level or at key points in the game), squad leaders, the antagonist's sergeants and lieutenants and the people in the bars and shops in the world the player character is exploring. By their very nature these characters frequently offer opportunities for the player to interact with them – to obtain items or information that enables the player to continue, or to confront the obstacle that the character represents.

If all that your significant enemy character says during this confrontation is something along the lines of, 'Prepare to die, vile human!' then it is likely to elicit inward groans from the player, particularly if the style of voice recording is also a little cheesy. If there really is no opportunity to have a significant, non-clichéd scene with the character, then it is better to have no scene at all and simply get on with the action.

Character interactions will often provide the player character with information or items they need. Story exposition, too, can be revealed through such interactive scenes. When the amount of information or exposition is large in such interactive scenes it is a fine balance between a dynamic interchange and running the risk of losing the player's interest. For instance, if the story exposition is too long, then you will probably need to split it up between a variety of characters or have the same character give the information over a number of encounters. Perhaps the character being questioned has to do more research to discover further information, or he could be wary of divulging everything because he does not trust the player character. Winning his trust could be part of the gameplay, as could finding items to trade for each piece of information or exposition. Whether that fits with the main gameplay ideas is another matter and would have to be discussed with the design team.

Too regularly I have seen games where the player character interacts with another to get an item and the other character hands it over as if the two of them are best buddies. Where is the dramatic tension? Where is the conflict? Could this be why so many players hate dialogue scenes – because they can seem to have no purpose and fail to keep the player's interest?

Tension and drama are created through something known as the expectation gap. This is the difference between what a character expects and what actually happens. What the player expects to happen and what actually does is the expectation gap. For example, a character may expect to open a door, but finds it locked; he then talks to the woman who has the pass key but she refuses to let him in and calls her boss who throws the character out.

In a game, because of its interactive nature, the expectation gap goes beyond that of the characters and extends to the player. Finding ways to overcome the expectation gap is the gameplay.

In game scenes where there is no expectation gap there will be no drama and so the scenes will come over as dull, particularly if all they exist for is to provide exposition. Without the conflict within scenes, the dialogue will never be made to shine.

Not all games are high drama, but even in a comedy game, say, much of the humour can be derived from the expectation gap by providing unexpected comedy moments. These can come from two characters completely at odds with one another because of a misunderstanding or because each has their own agenda and has no interest in what the other has to say.

Whether it is high drama, comedy or a mixture, making the expectation gap work can only be done when a writer knows the characters well.

The sidekick

A sidekick is a character in the game who the player character can rely on to help out during the game. This could be someone back at base that the player can call up on the radio and ask for assistance or a game-controlled character who battles beside the player character and follows them around the game world, taking a lead from the actions of the player.

Sidekicks are usually aligned closely to the player character because of this need to offer supportive gameplay, but this means that care has to be taken that the character does not simply become a clone of the player character. It could be that there are few opportunities for the two characters to interact with one another in any significant manner, but when they do there should be some tension or drama that fires up those scenes in a dynamic way.

Sidekick characters should always have their own, reasonably detailed back story. Even if only a fraction of this comes through during the game, you know it is there to draw on to create vibrant scenes. Back stories, along with good character profiles, are an important part of defining all the significant characters in the game.

There may be restrictions in what you write in the scenes between the player character and the sidekick, brought about by the needs of the gameplay roles of the two characters. It is going to be difficult to write scenes where the two of them fall out if they then have to rely on each other when battling the invading aliens during the next gameplay level.

In some games, sidekick characters are only available for some of the game to offer variety, or a series of sidekicks may come and go for the same reason. In these cases, the gameplay value overrides the story value and when a series of sidekicks are used they may be relatively superficial. These characters can still be well-developed, but should only be taken to a degree of development that matches their role in the game. There is no point creating a character of such depth that it would take hours to get to know them if they are only going to be on screen for half an hour, particularly if the player is occupied with action-driven gameplay.

The best scenario is when the same sidekick is used throughout and there is the opportunity to develop a sub plot that follows them. Why is this character working/fighting/investigating alongside the player character? What are their differences? What is the sidekick's motivation and how does this fit or conflict with the player character's motivation? Do they get on or is it just a job of work? Is there any special chemistry between the two characters – any sexual tension or same gender bonding? Will the sidekick change as a result of helping the player character or will the opposite occur? Will they both be forced to examine their lives as a result of the relationship and progress as characters throughout the length of the game?

The player character

Creating a strong player character for a game is not easy. A combination of the character's looks, the gameplay capabilities, the story role and the player's perceptions all have a bearing on how well the character will work. The player, in particular, is an important factor and because no two players are the same the way they view the player character is going to be different. Preventing characters from falling into cliché can be quite a challenge.

The expectations that arise from a game's genre should be an important consideration when developing the player character, for if the very thing players have most contact with fails to deliver what they expect, the likelihood is that the success will be severely limited. Much of this, of course, is a gameplay issue, but it is important that the character's defined personality traits do not contradict their gameplay properties. A confirmed pacifist is not going to be appropriate within the high action demands of a first person shooter, for example.

The motivations and objectives of the character should never be at odds with those of the player, though if the character and story have been developed well within the framework of the game's genre this is unlikely to be the case. It does not hurt, though, to have these objectives stated plainly from time to time as a reminder to the player that he is in synch with the player character.

The relationship between the player and the player character can vary in its closeness. With traditional point-and-click games, the direct connection is probably as tenuous as it can get without it being a game where there is no specific player character. In this type of game the interface is such that the player is effectively asking the character to move to a certain position or to interact with an object in the location, then waiting for the outcome. Successful characters in this type of game rely on creating empathy within the player in a similar manner to main characters in a film or a book.

In games where the interface allows the player to have more direct control over the movement of the character, the connection of the player with her on-screen avatar is much more substantial and the feeling that the player character is an extension of herself begins to take place.

At the extreme end of this spectrum are the first person games where the player character is never seen, even in cut scenes, and never speaks. The sole intention is to give players the impression that they have become that character, or even that it is the players themselves who have entered the game world and are directly fighting the opponents or solving the mysteries. In this situation the development of the character may be non-existent for the simple reason that to do so would break that feeling of direct connection.

In some respects, the members of the player's team in a squad-based game can be regarded as extensions of the player character. In many role-playing games, too, the player builds a team of characters and although they can often be interacted with, they are also an extension of the player character as the player can use their gameplay skills and abilities should the situation require it, but such characters are not sidekicks. Quite regularly, the player is given the opportunity to switch between the characters in a team and the currently 'active' character then becomes the player character.

Often in these situations, the story events and gameplay objectives are treated as if applicable to the whole team. The team becomes a kind of super-character with combined abilities and experiences. Though the story may single out one of the characters as the main protagonist, which character is being controlled does not affect the outcome of conversations or the information that is given by other characters the team will meet.

Point of view

The point of view of the player character is always different from that of the player, even in games where this difference is minimised as much as possible. A player, no matter how good the immersive experience, is always aware that he sits in front of a screen using some kind of interface device. The character always inhabits the game world.

When telling a story in traditional media, the writer must create it with a particular viewpoint in mind. This can be from the perspective of the main protagonist or from that of a dispassionate narrator. The viewpoint could also change as the story switches between a number of main characters.

In a game, the viewpoint should always be that of the player character, because the connection the player makes will be strengthened if this is the case. For that character's part of the story and gameplay, looking at things from their viewpoint will make it far more personal and so increase the empathy of the player to the character they are playing.

There are times when the viewpoint of the game and the main character is a little at odds with that of the player. This often occurs when the character has died and the player loads a previously saved game from a little earlier – the player will now know what is about to happen, even if the character is blissfully unaware. For many this is just part of the landscape of playing games and the player effectively knows how to set his mind so that he can continue without it spoiling his enjoyment. For games with a strong story, particularly those with an investigative gameplay element, there are occasions when the player knows things that the character does not. It is in these situations where the need to replay dialogue scenes that the player has already experienced should be minimised by giving the player the opportunity to skip them in some way.

Maintaining a connection between gameplay and story objectives also helps to keep the player's and character's viewpoint as close as possible. The player will always be thinking in terms of gameplay, whereas the character is not actually playing a game, but living through the experience of the game world. She will be 'thinking' about the events of the story and how they are affecting her.

There are some games in which the player character is changed for a level or two. It could be that the main character and the sidekick split up and the player is given the opportunity to play one and then the other in different parts of the game world. When the two characters join forces again, later in the game, the player obviously knows a lot more than either character. To consolidate this knowledge and help bring the viewpoints closer together once more, a summary scene could be very useful. However, I do not mean that each character launches into a long description of what they did while separated, but that you have something where the scene fades up and it appears that the player is just catching the tail end of the conversation. Something like, '... I was lucky to escape the landslide without serious injury'. In the player's mind, the two of them now know as much as each other and as much as the player.

Character profiles

The vast majority of game projects involve a large number of people filling a variety of roles. To ensure that everyone sees the characters in the same way, it is incredibly useful to create a series of character profiles, particularly for the main characters. It can be months between defining a character and it being modelled, animated or placed into the game – without clearly-defined profiles there would be a good chance that the initial vision for the characters will be lost.

Each profile should have input from the art department (concept sketch) and the design department (gameplay features, artificial intelligence definitions) and the amount of detail the writer contributes is dependent on the importance of the character to the story. The scope of the story will also help define the amount of work required. If the story is relatively superficial or the character interactions have little depth, it is probably wasteful to create highly-detailed profiles. Look at the way the characters are going to be used in the game, the expected depth of the dialogue scenes and tailor your character profiles accordingly, picking out the features that are relevant.

Sometimes, thinking through the features of each character forces you to approach them with more care and attention to detail. Even something simple, like deciding on their favourite colour, forces you to put yourself in the mind of the character.

The appendix contains a typical character profile. The number of these categories I use is governed by the depth of the story and how the characters will interact in the game. How you modify it to your own needs is down to how best it meets your own way of approaching your characters and how it fits with the design team's expectations. There is no single way of developing a character profile, but it is valuable that you do so in some manner.

Conflict and motivation

One thing that may be apparent – assuming you have read the previous chapters – is that it is very difficult to separate out the many aspects of writing for video games. If the whole is to be greater than the sum of its parts these aspects will overlap and interweave with one another in a rich and diverse manner. For instance, the player's choices have an impact on the story, which can affect the characters and in turn affect the gameplay and modify further player actions. This is a good thing, because if you were able to separate each of these things from one another completely it is unlikely that you would be able to treat game writing with the cohesion it needs. Conflict and motivation have already been touched on in earlier chapters, but here we will look a little more closely.

Gameplay is all about conflict. Without the conflict inherent in games, there would be no gameplay and no sense of achievement when the player overcomes his objectives, whether they are posed by playing against another person or against the game itself. This is not a new phenomenon that has arisen since the advent of video games, but one that has been around since the invention of games in any form. In games like chess and backgammon, which have been around for hundreds of years, the conflict is set by your opponent's moves as they attempt to remain one step ahead of you. In turn, you must try to resolve that conflict by using your own moves to throw it back on your opponent so that they are now the one with conflict to resolve.

Balance is very important when defining gameplay conflict. If two chess players are very mismatched, it is likely that neither will enjoy playing against the other. For the better player, the game will offer no challenge because it is too easy to beat their opponent. For the weaker player, the game is too difficult and because they know that they cannot win it becomes a pointless exercise. Both these players lack the motivation to play a game of chess together.

What the above shows is that, unlike more traditional media, conflict and motivation in games are not restricted to the events, characters and situations within the games themselves, but also extend to the players.

Gameplay styles reflect the different conflict and motivational needs of their players. For many fans of first person shooters, the buzz they get from working their way through the game's levels, blasting everything in sight can be all the motivation they need. For others a more cerebral challenge motivates them to play and they will look for games that deliver this more thoughtful style of gameplay.

Whatever the gameplay style, the writing and design must again fit together well if the motivation and conflict in the story is to complement that of the gameplay and keep the player's interest high enough to play through the game.

How this parallel motivation might be developed is shown in the following example. The player has been progressing through the game with a sidekick fighting alongside the player character. At the end of a particular level the sidekick is captured, which not only changes the character's motivation in a story sense, but also has an effect on the gameplay – the player must now play the game without the sidekick until a rescue can be made. The gameplay has unexpected variety and offers a different challenge because of the increased conflict the character – and player – now has to endure. The motivation of the player to get the sidekick character back as a gameplay aid matches the motivation of the player character in the story to rescue the partner.

Looking at the above, we can see that conflict has been created by an expectation gap (the player did not expect the sidekick to be captured), which is at the heart of how successfully it works. The impact of the expectation gap on the player will depend very much on how he is identifying with the main character and, in this case, the sidekick. If there is no empathy for the characters and no interest in their relationship, their motivations will mean nothing and the player just moves onto the next level with interest only in the gameplay. Again, it comes down to how well the many aspects are woven into a complete whole.

Conflict through the use of the expectation gap must be balanced with a series of rewards for the player. Do not lose sight of the fact that the game must feel and play like a game. Without rewards to give a regular sense of achievement, the motivation to keep playing the game will dwindle, no matter how dramatic the plot or character interactions are. In the example above, if the sidekick is captured and the player is left with a feeling of failure with no idea of what to do next, then they may feel there is no reason to continue with the game, particularly if it is the latest in a series of setbacks.

The loss of a main character is clearly a major setback, but if the player discovers the antagonist's operational base in the process, the reward not only keeps the player motivated, but the story also moves forward. The player and character are both motivated because they have new goals and a means of moving towards achieving them. Variations on this example could be that the player is forced to sacrifice the sidekick in order to get the information, but will attempt a rescue when it has been delivered. Or the player can plant a tracking device on the vehicle in which the sidekick is being abducted. Or it could all be a ruse and the sidekick being taken is part of a larger plan which ends in the sidekick being revealed as the true antagonist. The possibilities are endless.

A few years ago the ideas behind game design and development were often different from what they are now. At one time some designers thought that the purpose of setting puzzles and other gameplay obstacles was to stump the player. A kind of 'they'll never get past this' attitude existed as if the designer was in a battle to defeat the player. Thankfully this approach to design has pretty much died out and the idea of obstacles is to provide challenges, but ones that can be overcome through the use of the gameplay features the player has at her disposal. Gameplay is seen as a partnership between the developers and the players to provide an enjoyable balance between conflict and reward without the player becoming frustrated by a seemingly impossible barrier.

Story and character interaction are an important part of this blend and the player must not become frustrated because they are unable to keep track of the plot or interacting with certain characters makes no sense. When playing a game in a different order confuses the plot or fails to produce the right character responses because the triggers have not been tripped, then you need to put in a lot of work to correct this oversight. Be aware of these potential pitfalls or you run the risk of obstructing the player's progress and taking away their motivation to play.

You should also ensure that the use of contradiction is not replacing true character conflict. Having two characters simply arguing back and forth serves no purpose other than to exasperate the player with a scene that is going nowhere. Game players can be very impatient with anything that halts or slows their playing of the game, so any scene – interactive or otherwise – must create a feeling of interest for the player.

As is normal in other media, scenes should be constructed so that there is a difference when it ends. The characters will have clear intentions when they enter a scene, and what happens during the scene and how it ends is dependent on the motivations of the characters and how they are able to use them to their advantage during any conflict.

For a story that is entirely linear, the writer is able to take control of all the scenes in the game. However, when the story is non-linear and the scenes are interactive, the dialogue choices of the player or the order in which the game is played could have an outcome on how each scene pans out. Creating conflict and drama when the player is in control is more difficult, but with very careful planning it can be made to work in your favour if the characters react to the different choices of the player in a way that builds a richness and makes the characters seem fully rounded. If the player can talk to a character about a cat, a missile and a burger, you must write each small section of the scene so it makes sense if one player chooses to ask about the missile, then the burger, then the cat while another player talks about them in a different order. Writing in this way at the same time as keeping the scene interesting can be a real challenge.

Motivation of the antagonist is important, too. Without it, they will come across as shallow and you run the risk that the climax of the game will fall flat because there has been no proper setup for this character. It is important to give the impression that the villain is working on their plans even when they are not on screen, otherwise it can feel like they have simply been waiting for the player character to turn up each time they encounter one another.

We see too frequently in games, situations where the player character – and the player – know nothing of the villain until the end of the game when they are suddenly presented in the final boss battle. This hardly ever works if you want to create a strong story, so consult with the design team to find ways to set up the main antagonist as early in the game as possible.

It could be that discovering the antagonist is part of the story and gameplay, but that does not mean the character cannot be setup before this discovery. He could be one of the other characters in the game that, once the facts fall into place, makes perfect sense. You could show the mind of the antagonist by the trail of victims they leave behind and the clues the player must piece together to reach a final showdown.

Just as conflict and reward must be balanced, pacing is also an important element helping create that balance. Unfortunately, because of the interactivity involved, controlling the pacing can be a problem. How often plot information is revealed or how regularly the player character gets to speak to other characters depends on the route taken through the game and how quickly the player moves along that route.

There are always ways to maximise the pacing of a game and there will always be some optimal path that most players will follow. Build the pacing to play to the strengths of this common path, but think through the other possibilities to ensure that there is no over-concentration of story and character conflict in small areas.

Dialogue and logic

Dialogue is the most direct connection players have with the craft of the writer and it regularly comes under the most scrutiny. The style and quality of the dialogue will have a strong bearing on the overall impression the game has on the player, particularly how it sits with the characters, the settings and the style of play.

Just as everyone has an opinion on art, it seems they also have strong views on dialogue and feel they know when something is badly written or poorly acted. Even dialogue that is generally regarded as being of a high quality will have its detractors; such is the diversity of tastes and opinions. Gaming can often bring out the most extreme views from players and dialogue is a regular target for them.

The dialogue in adventures – a genre well known for the large amounts they contain – can still elicit very mixed views from the players. When *Broken Sword* was released in 1996 many people loved the huge amount of rich dialogue, yet there were many others who felt that there was simply too much and that at times it got in the way of playing the game. When the sequel was released in 1997, the amount of dialogue was drastically smaller, which won over a number of people, yet others felt the depth of the original was lacking.

Clearly, finding the right balance for the amount of dialogue is as important as the quality, but that balance will be one which will differ from genre to genre and even from game to game.

The play pauses

Each time characters speak means the player stops playing the game, however momentarily. Therefore, creating scenes which are direct and to the point minimise this non-interactive time, particularly if the scene is part of a high action game.

When games first moved onto storage formats with much larger capacity (CD-Rom, for instance), many games used pre-rendered full-motion video (FMV) sequences to portray scenes in which important characters spoke to one another and vital story information was revealed to the players, who were initially excited by these sequences. Often a game was judged in part on the quality of its FMV. As the standard of games has improved and FMV is no longer the special feature it was, such scenes must really pay their way if they are to be included in the game's development. Long, lingering establishing shots and fly-throughs are becoming a thing of the past and FMV must be treated as if you were writing the tightest of film sequences.

Today, most pre-rendered sequences have been replaced by ones that use the game's engine, which enables a visual consistency throughout. However, the principle of the FMV sequence is retained and the player has no interaction while it plays through. Such sequences need to deliver dramatic, relevant exchanges which pass on information in a way that does not lose the player's interest.

The makers of *Half-Life 2* took a different approach to the problem of the loss of player control and rarely had a point in the game where control was ever taken away. When dialogue scenes were triggered, the player was still able to move around and interact with other objects or even walk away from the conversation altogether. The down side of this, though, meant that it was very easy to miss important story information. If the player was interacting with something else at the time, they would not necessarily be paying attention to the unfolding dialogue and this can give the player the impression that such scenes and the information they impart are of little importance. The onus was on the player to get the best they could out of the experience and it was this aspect which appealed to many players – that their control was maximised within the style of game and the unfolding story.

The usual alternative is to ensure that dialogue scenes are delivered to the player in full, but in a way that is controlled by the player by making the dialogue as interactive as possible. The player chooses which topics to discuss in which order and when to exit a conversation. They should also be given the opportunity to return to the character and resume the conversation at any time. The control given to the player is a very different style to that in *Half-Life 2*, but one which maximises the interactivity of games which rely on gathering information by talking to others.

While the writer and design team may work together to define how dialogue scenes are presented to the player, whatever solution is chosen will rarely suit everyone's tastes. Trying to please everyone could lead to the creation of a system that becomes over-complicated, which in turn could lead to a lot more work when writing dialogue to match the system.

Dialogue systems

What is a dialogue system? This is the part of the game engine which is required to run the dialogue and the tools needed to implement the scripts in a manner that the engine can use. It may not appear to be very complicated to play some dialogue or display text on screen, but the reality is that with increasing sophistication of games, dialogue scripts can be extremely intricate. Of course, this will depend on the complexity of the game's dialogue requirements – a simple voice-over here and there will not require the complexity of a deeply-involved RPG – but showing how a script might work in different circumstances will create a clearer picture of how you might approach the writing.

Nearly all development studios create their own proprietary script systems because they need something that matches the requirements of the gameplay and even if they look similar on the outside, how they work and how the scripts are prepared can be very different. This in turn means that, unlike film, TV or radio, there is no standard format for script layout. Although games are a visual medium, the script layout of TV and film screenplays are less applicable and the format of radio scripts or even stage play scripts are closer to a 'standard' because they are easier to incorporate into the game and at recording time will offer fewer page turns as more dialogue lines can be placed on each page.

So what might a game script look like? To show how this might look, I will create a basic script and go through the process of how it might be turned into a game script. For the sake of argument, let's assume we have a scene where the player character (Edwards) is investigating a murder. He talks to a potential witness (Wilks) and the scene could begin as follows.

Edwards:	I heard that you witnessed the shooting.
Wilks:	That so?
Edwards:	Just tell me what happened!
Wilks:	Get lost! I didn't see nothing!

For many games, this may be all that is needed from the scene. The dialogue is not interactive and simply plays out when triggered by the player in some way. So how does the above transfer into a game script? It can depend on how writer-friendly the scripting system is and the following shows how this might look in a system that strives to be as readable and user-friendly as possible.

```
if(wilks_shooting == false)
{
    Edwards:        I heard that you witnessed the shooting.
    Wilks:          That so?
    Edwards:        Just tell me what happened!
    Wilks:          Get lost! I didn't see nothing!
    wilks_shooting = true;
}
```

Dialogue scripts are often brought into the game using a high-level type of programming language which enables the writer to check for the right conditions to trigger the dialogue. In the above scene, we only want these lines to be spoken once or the suspension of disbelief will be destroyed if we re-run the same scene every time the player character interacts with Wilks. A variable is used, wilks_shooting, for which the engine checks and plays the scene if the variable is false. Once the scene lines have played out, the variable is set to true and the scene will never be triggered again.

In a less user-friendly system the scripts are a little less readable and take more time to implement from the initial dialogue, but with care and attention can quickly become almost second nature. A script done in this fashion may look something like the following.

```
if(wilks_shooting == false)
{
    character_speak(Edwards, 'I heard that you witnessed the shooting.');
    character_speak(Wilks, 'That so?');
    character_speak(Edwards, 'Just tell me what happened!');
    character_speak(Wilks, 'Get lost! I didn't see nothing!');
    wilks_shooting = true;
}
```

In this example, character_speak(A, B); is a scripting function with two parameters. A is the character object and B is the text string which A speaks. In the previous example, a similar function would also exist, but will have been hidden from the user in the toolset by those developing the system to make it more user-friendly. For the remainder of this discussion I will use the simpler style for clarity, but neither of these styles is close to a standard and some tools do away with this kind of approach altogether, but may suffer in readability, depending on how they have approached the issue.

If the scripting system has a feature which allows you to place comments into the scripts, you should always take advantage of this and put in comments at every opportunity. Comments are useful as a reminder to you, but also help others who are working on the scripts – implementing puzzle logic or facial animation for the characters, say. They can also be a useful help in the studio when recording the voices as the comments will be able to put each small section of script into context. If we take our script and put in a few suitable comments it may look like the following.

```
if(wilks_shooting == false)
{
   //Edwards launches straight into his first line of questioning with no pre-
   amble
   Edwards:        I heard that you witnessed the shooting.
   Wilks:          That so?     //Nervous, but puts on brave face
   Edwards:        Just tell me what happened!
   Wilks:          Get lost! I didn't see nothing!
   //Edwards looks angry – he should have handled it better
   wilks_shooting = true;
}
```

The double slash (//) and the italics is a common style of putting comments into code. As the game scripts are a simple form of code, it helps if such conventions are maintained. Because comments are never compiled into the final code, even a comment written at the end of a line will not appear on screen if the dialogue lines are displayed as subtitles.

Comments are also useful for keeping track of any logic that might be a part of the dialogue scripts. Sometimes the dialogue itself will not tell you where the scene snippet lies in the bigger picture and a comment is necessary. If the game allows the player the freedom to talk to characters repeatedly, it is sometimes necessary to create a generic line or two so that there is a response of some kind. The logic is pretty simply and, with comments, may go something like this:

```
if(wilks_shooting == false) //Talk about shooting for first time
{
   //Edwards launches straight into his first line of questioning with no pre-
   amble
```

```
    Edwards:        I heard that you witnessed the shooting.
    Wilks:          That so?      //Nervous, but puts on brave face
    Edwards:        Just tell me what happened!
    Wilks:          Get lost! I didn't see nothing!
    //Edwards looks angry – he should have handled it better
    wilks_shooting = true;
}
else                //Subsequent times
{
    Edwards:        Tell me about the shooting!
    Wilks:          I got nothing to say.
}
```

The problem we have created here is that talking with Wilks appears to be a dead end unless the player can find some way to leverage the information from him. It could be that another character in the game offers the opportunity to do this and gives the player information which helps resolve this conflict. Because you want to create scenes that respond to the information the player character holds, make them as naturalistic as possible and minimise the need to repeat sections of dialogue, the level of complexity can begin to rise swiftly. Expanding on our example, the player gets the leverage he needs from a character called Johnny. Because he could have talked to Johnny either before Wilks or after he talked to Wilks, this must be taken into account when thinking through the logic and constructing the scene.

```
Edwards:    Hey, Wilks.           //Greeting used every time
if(wilks_johnny == false)         //Not spoken to Wilks about Johnny
{
    if(wilks_shooting == false)      //Talk about shooting for first time
    {
        Wilks:          What's up?
        Edwards:        I heard that you witnessed the shooting.
        Wilks:          That so?      //Nervous, but puts on brave face
        Edwards:        Just tell me what happened!
        Wilks:          Get lost! I didn't see nothing!
        //Edwards looks angry – he should have handled it better
        wilks_shooting = true;
    }
```

```
        else                //Subsequent times
        {
                Wilks:              I got nothing to say.
        }
        if(johnny_shooting == true)     //Talked to Johnny about shooting
        {
                Edwards:            Your friend Johnny saw you with the body.
                Wilks:              That junkie ain't fingering me!
                Edwards:            It's not looking good, man.
                //Wilks thinks, weighing his options
                Wilks:              Listen, all I saw was a guy in a leather jacket
                                    running away. The woman was already dead.
                //Leather jacket is key info
                Edwards:            Thanks.
                wilks_johnny = true;
        }
}
else
{
        Wilks:              Get lost, will you?
}
```

The main part of the scene will play out, in one way or another, each time the player interacts with Wilks until the two have talked about Johnny, at which point the variable, wilks_johnny will be set. After that, the only response is for Wilks to say 'Get lost, will you?' each time.

If the player has talked to Johnny before Wilks (setting the variable, johnny_shooting) the two longer sections of dialogue will play one after another, so they must work together as if constructed as one scene. However, it could be that the player only triggers the first part if Johnny has not been spoken to, so this also needs to stand on its own. The player then has to find Johnny and talk to him in order to trigger the second part of the above scene when interacting again, which means that the second of the longer sections must also play out independently. Adjusting the first of Wilks's generic responses and putting in an initial 'Hey, Wilks.' helps smooth any potential awkwardness.

It may be that as a writer you are protected from such logic scripting and that the design team will implement such logic into the game, but by being aware of how it works will enable you to write your scripts in a manner which makes it much easier to implement and reduces the need for changes and additions once the logic is applied. Your script could look like the following:

Scene – Edwards talks to Wilks

Edwards: Hey, Wilks. *//Greeting used every time*

First Time

Wilks: What's up?

Edwards: I heard that you witnessed the shooting.

Wilks: That so? *//Nervous, but puts on brave face*

Edwards: Just tell me what happened!

Wilks: Get lost! I didn't see nothing!

//Edwards looks angry – he should have handled it better

Other times (not talked to Wilks about Johnny)

Wilks: I got nothing to say.

Edwards has talked to Johnny

Edwards: Your friend Johnny saw you with the body.

Wilks: That junkie ain't fingering me!

Edwards: It's not looking good, man.

//Wilks thinks, weighing his options

Wilks: Listen, all I saw was a guy in a leather jacket running away. The woman was already dead.

//Leather jacket is key info

Edwards: Thanks.

Repeated response line when all information is obtained

Wilks: Get lost, will you?

If the structure of the above is not clear enough to the design team, or whoever is implementing your scripts into the game, you should be prepared to elaborate, either by talking to the people involved or putting further instructions into the script itself, whatever works best for everyone involved.

The above example could be further complicated if Edwards has found out other information about Wilks. If he has already talked to Sally and found out that Wilks is a drug dealer, say, it could affect his attitude towards him, in which case you may need to create a whole new version of the above scene which reflects this. There is a danger that this approach can get out of hand, so wherever possible keep it simple and try to create lines which suit most eventualities. If necessary, put an extra line or two into the scene – wrapped up in suitable logic – which could change the flavour without the need for a complete re-write.

Holding all this in your head as you write can be pretty daunting, so it's sometimes a good idea to develop the structure of the scenes before writing the dialogue within it. Ask yourself plenty of questions as part of this process (What if Edwards has spoken to Sally? What if Edwards has spoken to Johnny?). Be sure to check with the design department that what you are developing is correct. There is no point writing part of the scene which assumes Edwards has spoken to Sally if the player does not get the opportunity to meet her until much later in the game.

The actual logic structure I have used here is only one possibility and will vary from project to project and how interactive the dialogue scenes are going to be. How the design and implementation teams will set out the logic and put it into the game must be understood by the writer as it may affect the way the dialogue scenes are written.

The scripts for dialogue scenes can become littered with many other functions, which will depend on the sophistication that the engine can deliver. These functions could be to play an animation (the character scratches their head while speaking), to change a character's facial expression (suddenly looks angry as a result of what the other character said), to move the character to another position (walks over to the door), to change the camera position, or numerous other possibilities. This can have the effect of making the scripts much harder to read, but many scripting tools have the ability to set up colour-coding. By having functions appear in one colour, variables in another and dialogue in a third, say, the scripts become a lot more readable again and you can learn to ignore what is irrelevant to you.

Often, the dialogue can be exported from a complex script and into a separate file, where it can be edited much more easily and then imported back into the toolset once more. This usually means that the dialogue is exported into a spreadsheet, which is actually very useful when the dialogue is localised into other languages.

If you do not already know how a spreadsheet works, it may be worth your while gaining some degree of familiarity, though you do not need to become an expert. When editing dialogue in a spreadsheet, it can be better to create a new column for the changes, which makes them easier to spot by anyone reading the spreadsheet. However, make sure that you check the preferred method with the design team.

The player chooses

Some dialogue systems offer the player the option to choose the subject they are going to talk about. This may be presented as a list of questions that the player is able to choose from or as a series of icons that represent the subject in a more conceptual way. In the example above, and after the initial greeting, the choices presented to the player could be 'The Shooting', 'Johnny' and 'Sally'. Whichever the player chooses, a segment of the scene is triggered which relates to that choice.

Because the subjects are dependent on the player having talked to other characters or gaining other knowledge, the choices should not be presented to the player if the right conditions are not met. This, and the fact that the player can choose which order to talk about the subjects, means that the dialogue must be written in a way that takes this into account and additional logic variable must be applied to ensure it hangs together well.

Some subject choices may be completely independent of others so do not need complex logic to handle them, but if a character is pivotal to the plot this may throw up a lot of subject choices that interweave intricately. Choices may disappear when discussed, as is normal, but then reappear later when the player character knows more information. The dialogue system and tools need to be able to handle this sophistication if this is the type of dialogue gameplay the project requires.

The non-speaking main character

Because some games offer the player the opportunity to define or customise the player character to a high degree, they throw up a different set of requirements. It is difficult to write for a character that could be any of a huge number of possible combinations. Features the player can regularly choose from include gender, race, profession, abilities, age, background and their good/evil disposition. Depending on how many choices are available in each category, the number of variations could be enormous and to write and record dialogue for each one is an impossible task – many of these games will already have high dialogue content before adding these factors.

The only way to handle the dialogue in these situations is for the player character's dialogue to be very minimal, neutral and be able to suit the range of possibilities. It also means that the lines of the player character are not recorded. In some games none of the dialogue is recorded, which at least gives a consistency to the dialogue.

Developers may record the dialogue of the non-player characters because they tend to be non-definable, but this can lead to an issue where conversations appear unbalanced with only one of the two characters speaking.

The development studio, Bioware, handled this very well in their Star Wars RPG title, *Knights of the Old Republic.* Whenever it was the player character's turn to speak, the player was presented with a list of questions to ask or lines to say. The action of choosing one of the lines from the list worked as if the player character had already spoken the line and the other character would speak their response immediately. Like many other aspects of gameplay, when handled well the player quickly accepts this kind of stylisation as part of the game world, but care has to be taken that the right balance is maintained and nothing destroys the player's suspension of disbelief.

The down-side to this approach is that it is so much harder to create dynamic dialogue scenes when the game pauses for the player's response each time and we never hear full to-and-fro exchanges. To minimise the need for the player to keep making choices, the speech lines of the other characters can be made longer, often written in a way that foresees any follow-up questions. However, without care they can come across as a series of short monologues that the player triggers rather than true dialogue exchanges. Finding a good balance can be a lot of hard work and a feeling for how this kind of game is played and the way the scenes unfold is vital.

Misleading or guiding

Style, quality and quantity of dialogue vary enormously from game to game depending on each one's requirements. The writer, though, must be very careful that the dialogue does not mislead the player unintentionally, which could mean that he or she becomes frustrated trying to achieve something that is not actually possible.

I was developing the prototype for a game and at one point there was a large log with which the player could interact. It was impossible for the character to pick up the log, so I wrote a voiceover line which I believed would convey this: 'I'm not strong enough to pick this up.'

Unfortunately, when we did some focus testing, a significant portion of the players took this to mean that they had to find a character strong enough to pick it up or to find a potion or other item that would give the character the strength to do so. For these players, this line of reasoning appeared to be confirmed when the main character met a stronger character a little later in the game.

Clearly, this misleading of the player was unintentional, but highlights how important it is to look at your dialogue in the context of playing the game and trying to see it from a player's perspective. If you do not catch such instances, the player could waste a lot of time trying to find a way to do something that the game will not allow.

The flip side of this coin is the way that such lines can be used to guide the player. If it was important that the player picked up the log, the original line I came up with would have been a subtle clue that a stronger character was needed to do this. For some types of game this line may be too subtle and a more blatant direction could be used: 'I should find someone to help me with this log.'

The advantage of the more subtle approach is that you can guide the player without being obvious and the gameplay becomes a little more of a challenge without being impossible.

Finding the right level of clues and guidance that fits the style of the game, the nature of the main character and requirements of the gameplay is something that must be worked out with the design team.

The writing team

Some games can have ten thousand lines of dialogue, or even more. As a typical movie contains about a thousand lines, it quickly becomes clear that these games have the equivalent dialogue of ten movies. Asking one writer to create scripts for ten movies would be a huge expectation and one that is unlikely to produce consistently high quality and still fit within the time-frame of the project's development. The only answer is to use a team of writers on the project who must each understand the way that gameplay, story and dialogue mesh together.

Whenever a team of writers work in this way, they must ensure that their writing styles match, but this is only part of the picture. Understanding the style of the logic structure that drives the dialogue scenes will enable the team to maintain a consistency in the way the game's scenes unfold.

Even with the best will in the world, maintaining consistency of writing style will be difficult without the writers constantly reviewing each other's work, so time must be allowed for this. Alternatively, it may be better to have one writer who acts as the script editor for the project. This person could be one of the writing team, or it could be a person who has been brought in specifically to fill that role once the scripts have been completed.

The script editor should always have the ability to look at the game and its dialogue as a complete entity which will guide the way they look at the fine details of the scenes and the lines they contain. The need for changes is probably going to be great as the editor tries to establish a consistent style, so if you are that editor make sure that enough time has been built into the schedule for you to complete the task satisfactorily.

Testing and polishing

Before the dialogue for a game can be recorded, it must be thoroughly tested to ensure that playing through does not throw up any discrepancies, oversights or logic bugs that could cause problems for players. If the game has a non-linear plot and character interactions that take place in any order, the dialogue testing can be quite complex and should be carried out by a professional team employed by the development studio. In the case of games that have a writing team, the writers should play each other's sections, which not only will help to catch the bugs but will also show up any inconsistency of style and help the script-editing process. Whatever the setup, the writers will need to fix dialogue bugs as quickly as possible.

Story oversights and lack of clarity are usually pretty easy to fix with a few additional lines or an additional scene or two. However, care should be taken that the fixes themselves do not cause other problems and should be thoroughly tested, too.

Logic bugs are potentially the biggest problem and can seriously affect the way that a game plays and the story unfolds if they are not caught and fixed accurately. Usually it is the implementation or design team that will hunt down and fix the bugs relating to dialogue scenes, but you may need to work with them if fixing the bug means re-structuring the scenes connected with it, editing lines or writing new scenes. This is where the act of putting comments into the scripts begins to pay off. You should know what each section of dialogue is meant to represent and how it is supposed to work. For instance, if you have been very thorough you may have put in comments to remind you where conditions were set to trigger the scene – in which scene the variable was set that the current scene is testing for.

Try to become part of the testing process yourself and play through the game – or your part of it – with an eye for whether the story and dialogue are working as they should. Care should be taken to play through it in as many different ways as you can. If you find that you play through in the same order each time, force yourself to take a different route. Does it still hang together? Is it just as enjoyable?

If the dialogue is not thoroughly tested and fixed before the actors enter the studio, the likelihood is that bugs or mistakes will be found later that could turn out to be very costly if actors have to return to the studio later. Publishers, as well as the console manufacturers, test the game very thoroughly when it is nearing completion, so even if there are suspect lines that you might be hoping to squeeze through, the expert testers will find it and fail the version.

A final note

Although my intention in this chapter is not to teach you how to write dialogue but how dialogue should be approached in relation to developing a game, one thing must be borne in mind at all times. Dialogue is meant to be spoken.

No matter how good it looks on the page or the computer screen, if it sounds clumsy when read out loud it is not going to work in the game. Get into the habit of reading your scenes out loud and acting them out if possible. Not only is it a very useful way of catching problems, it is also a great deal of fun. Because I work from an office in my home while my partner is at work, I am sure that my next door neighbours must wonder what is happening when I appear to be having dramatic arguments with myself.

Comedy

Just as in any other media, comedy in games is a very subjective thing. Lines or routines that will have some people in stitches will leave others very cold. We all have our favourite comedies on TV and can be quite surprised when we speak to others who do not share our love of those programmes. This subjectivity means it is very difficult to write comedy that will appeal to everyone and to try would probably lead to the humour feeling forced at best and may very well fail altogether.

The opportunities to write comedy games are much less common than they used to be. There has been a definite move towards games with a darker, more serious feel and such games dominate our perception of the current state of gaming. While there are many games that have funny, even hilarious, elements, the humour mostly tends to be physical and slapstick in nature, but without care, even these games can prove to be problematical.

Probably more than any other area of writing, making sure that the comedy works is vitally important to the way the game is received. Players may overlook weak lines in a more serious game, but if a joke falls flat it will stick in their memory because this is the heart of the comedy game. If the humour and jokes do not work, then you really do not have a comedy game at all.

The writer and designers must have similar senses of humour to create the game. Without this, there is always going to be conflict over what is funny or not. There must be a shared comedic vision for the game that parallels the gameplay vision and how the two of them go hand in hand through the design and development of the game. Is the comedy going to be slapstick or a more subtle brand of humour? Is the comedy going to be interactive in some way and if so how is this mechanism going to work? Are there visual jokes as well as dialogue humour? Try to plan as much as possible about the style of comedy and the way it is going to be presented to the player so that you are able to maintain consistency.

Team contributions

Working on a comedy game can be tremendous fun for the whole development team as the ideas are created, refined and implemented. However, this regularly leads to the situation where everyone thinks they have great ideas for jokes or comedy lines that should instantly be a part of the game. To prevent the development falling into chaos, such team contributions should be carefully controlled through brainstorming sessions.

The design team and the writer should be the ones who decide which suggestions are included and should be wary of any that are inappropriate. These can range from company in-jokes – which will clearly never work for the players – to silly suggestions that in the context of development are hilarious, but in the context of play probably will not make sense.

Some suggestions can be brilliant, of course, so keep an open mind in case one of the artists or programmers comes up with a real gem that gives a lot of mileage. However, these contributions should be handled with caution – even a brilliant suggestion may have to be excluded if the style of the humour does not fit the situation or the characters. You should always look at the possibility of taking good ideas and adapting them to the required style. Brainstorming can often be more valuable for the tangential ideas that are thrown up than for the initial ones.

Repetition

One of the biggest potential problems in developing game comedy is repetition. Most humour is based on delivering the unexpected, which does not always sit so well with games in which sections of play could be repeated before the player succeeds in overcoming the obstacle in question.

Any repeated dialogue can be very wearing on the player and generic lines must always be handled very carefully. When those generic lines are an attempt at humour they can suddenly become the biggest reason to hate the game. The line could be hilarious the first time the player hears it, but if it keeps cropping up throughout the game it can drive the player to the point of complete distraction. Not only can the player become extremely annoyed by the joke's repetition, the character will appear to be stupid and lacking in depth because they only have a handful of no-longer-funny lines. The character turns into the class outcast – the poor guy at school who constantly repeats a joke because it got him a laugh the first time.

Although this humour repetition can be easily prevented by ensuring that there are no jokes in any of the generic lines, other kinds of repetition should also be avoided. This is where the logic structure outlined in the previous chapter can really help. By wrapping any humorous exchanges in conditional checks you ensure that the player does not trigger the same section of dialogue more than once without starting the game again or restoring an earlier save game, both of which are outside the writer's control. Setting a simple true/false variable once the section has been run for the first time will prevent it from running again. Sometimes well-written comedy can give a game replay value, particularly if it can only ever be triggered once in each play through. Players have been known to replay games just so they can hear the dialogue again.

Logic structure can also be used to advantage even when the game is restarted, if you have the opportunity to do so. A series of conditionals could be set that each has a humorous exchange within it or a snappy one-line reply contained within each one. If the condition to be met is based on a variable that is set to a random number, each time the game is played from the beginning or from an earlier saved game the exchange is chosen randomly. This, of course, is more work for the writer, but when players and reviewers pick up on the extra detail you have employed – providing it works well – the response is very gratifying. It should also be noted that you should not need to do this for every comedy exchange throughout the game, but concentrate on the feature where it will have the most impact, particularly near the beginning or around areas where the player is likely to have to replay sections.

Surprise and risk

Because much comedy works best through surprise (a sudden, unexpected pie in the face or conversational lines that take an unforeseen turn) a writer sometimes needs to take risks to deliver the surprise necessary. After all, the world would be a poorer place if the Monty Python team had not taken risks with each sketch and film they produced. By its very nature, risk has its dangers and you must be prepared to be both loved and hated for taking risks.

Unfortunately, getting your risky comedy into the final game can be problematical if there are others who do not buy into what you are trying to do. Even people who love your scenes and laugh out loud may not be so happy about being a part of the risk. Not only will you have to convince the design team, but probably the game's director and producer, as well as the publisher's representatives. You therefore need to be clear about your intentions and convinced that it will deliver a laugh that is in proportion to the size of the risk taken. You need to have complete belief in your idea so that you can persuade others that it will work and to make it work in the game itself.

Interactive comedy

Games are an interactive medium, so it is easy to think that a humorous game must naturally have interactive comedy. However, this is not necessarily the case and the game may give comedy moments with which the player has no interaction while it is playing out.

Interactive comedy is that which only unfolds on input from the player. It could be something simple which delivers a slapstick moment or a more involved revealing during an interactive conversation.

An example of the simple slapstick could be the old 'Do not press this button' sign next to a big button. When the player presses the button (as they invariably will), the possibilities are enormous, but mostly used before – pie in the face, trapdoor opens, one-ton weight falls, etc. Alternatively, the player tries to interact with the button and the character says, 'I'd better not ...'. It could also be set up so that if the player persists and tries a few times the character gives in to temptation and presses it, delivering the slapstick moment. How long you hold off would need to be balanced by the expected humour of the delivery.

This side of interactive comedy is actually not centred on the element of surprise in a straightforward manner. If the player knows that a certain interactive action will result in something humorous, it must be delivered in a way that builds on the anticipation of the moment. In the game, *Beneath a Steel Sky*, there is a point where, to distract a guard, the player must drop a dog into a pond. The puzzle is quite intricate and involves luring the dog onto the end of a plank with some biscuits and then dropping a pile of bricks on the other end. However, rather than simply flipping the dog into the water, the see-saw effect catapulted him into the air like a rocket and he was off screen for a few seconds before coming back down and landing in the water with a huge splash. Not only had the puzzle delivered the humorous expectation, it had exceeded it with the extreme nature of the animation, created by an animator who bought into the whole joke. This shows the value of the writer working with the other members of the development team, particularly when the comedy relies on visual aspects.

Interactive comedy in dialogue scenes tends to work best when the character with which the player interacts has their own agenda. Though the player character may want to ask questions that are relevant to the game's plot or gameplay, the other character may only be interested in talking about themself and try to twist around the answers to the questions. So asking about the name of a nightclub could trigger them to reminisce about when they used to be a cabaret singer, say.

Clearly, such scenes have to be handled carefully or the agenda of the other character to talk about themself could just prove to be a distraction that annoys the player when trying to get on with the investigative gameplay. You have to balance this distraction with how you deliver the information the player needs to know.

As discussed in the previous chapter, the logic that wraps up the scene is very important and where comedy is written into the dialogue scenes you need to be more careful that the scene hangs together if the player has the option to vary the order in which the interactions take place. The comedy could fall completely flat if it relies on information that has not yet been revealed to the player. If such a variable nature exists in the character interactions, it may be best to ensure that any humour is self-contained within each segment, even if it advances a common or recurring theme.

Interactive comedy can also derive from how characters react to what other characters do or say. If you talk with Bill and then tell Jenny what he said, the reaction may be very humorous and may give a further chance for humour if the player then returns to Bill to talk about Jenny. In these situations, it is best if the characters are not too far apart, geographically, or the player character will be forced to trudge back and forth a long way to trigger the humour and it may take the edge off it.

Testing, feedback and brainstorming

Testing reaction to your comedy is important and the feedback on what works is vital. Sometimes a section of humour may seem right to you, but when people play through it they may find that it is not as funny as you think. This does not necessarily mean that the whole thing needs re-writing; it could be that a couple of tweaks here and there will change the whole thing around.

It can be difficult to create comedy on your own, particularly the tight, snappy variety. Brainstorming can be of immense value because it provides a situation where you can be as silly or outrageous as you like. Not only is brainstorming humorous ideas great fun, it can throw up ones that none of those participating would necessarily have thought of on their own. The brainstorming session can also be an excellent way to recharge your humour batteries and fire up your writing.

Many of the best comedy TV shows were created or developed by more than one writer and many long-running series will employ a large number of writers, often working in teams of two or more, because there is often more to be gained by writers firing off each other.

Even when you are the sole writer working on a game, it is useful to have brainstorming sessions with the design team who, at the very least, will act as a sounding board for your ideas. It is a way you can all have some fun at the same time as the designers are buying into your ideas and contributing some suggestions of their own.

The testing process can throw up a problem with repetition, but in a different sense to that mentioned earlier. When a game is played over and over again to test for bugs, even the best of comedy dialogue can become a little tiresome. This can lead you to worry that the comedy is not as funny as you first thought and while this may be true in some cases, you often have to rely on your initial judgement. Reworking comedy may improve the quality, but there is also a risk that you can over-work it and it then becomes forced.

The sense of fun

Delivery is also important to the success of the humour when the dialogue is recorded. In *Broken Sword – The Sleeping Dragon*, there was a minor character, a chef, which the player had to interact with in order to progress. Although his lines were well crafted and very funny, the performance by the actor was excellent and lifted the whole scene to a better comedic level.

Actors are also a very good final test of whether the comedy is working. If the actors are having fun as they work through the scenes, not only do you know you have got it right, the flavour and feeling of the final game will be improved as the exuberance comes through in their performances.

Not all games that contain humour are comedy games, but the hearts of those that are should beat with a true sense of fun. Trying to create a comedy game through dialogue or regular one-liners when the gameplay takes itself too seriously is likely to give the player the feeling of a game at odds with itself. The sense of fun should pervade the whole game so that the comedy feels natural and consistent with the other aspects of the game world.

The demographic of game players is much broader than it used to be, so it can be harder to target the audience with the right level of humour. One approach is to look at examples in other media that go beyond a single demographic and see what can be learned from them. *Wallace and Gromit*, *Toy Story*, *Calvin and Hobbes* and the Harry Potter books are all examples which have a broad appeal.

Along with a brilliant sense of fun, there is also a strong element of wonder – children love these fabulous worlds and the characters that populate them, and adults are taken back to times when they would enjoy tales of wonder and fun from their own childhoods.

Character-driven humour

Much of the strength of the above examples comes from the rich characterisation they contain and how a large part of the humour is driven by the characters. While there will always be jokes, we also laugh at the way the characters react to their situations and to one another. To create humour of this nature it is vital to know your characters and understand the way that they will react to a situation. If the characters react consistently their believability will be strengthened in a way which means that even when the humour does not make the players laugh out loud the suspension of disbelief is not broken and the player continues to be immersed in the game.

Even if your comedy game is not intended for such a broad demographic, character-driven humour is still important. Top sitcoms like *Black Adder*, *Porridge*, *Frasier* and *Friends* have a strength and long-lasting appeal because of their character-driven nature. Regular repeats are often worth watching because they rely less on blatant jokes, which can become stale if repeated too often, and build characters we care about and laugh with rather than at.

Comedy games have a long way to go before they can develop an appeal that rivals that of the top sitcoms, but if we can create strong humour that interweaves with the gameplay, it could be that replay value of such games will be due, in no small part, to this comedy. Just as many people love to regularly watch their favourite sitcoms on DVD.

Licenses

Games based on licenses from other media are a major part of the game industry as owners of various intellectual properties attempt to maximise the revenues from their investment. While there are many who see it as an erosion of original game development, there is clearly a huge market for such games, if current game sales are an indication. It is also not a new phenomenon – licensed games have their origins in the early days of the industry – but the number and scale of licensed games has increased enormously of late and it would appear that the trend will continue.

Very few writers get the opportunity to create and develop a game based on their own intellectual property, which means that for most of you there is little fundamental difference in working on a licensed game project or working on the developer's own property. You will use your skills and experience to do the best you can within the scope of the project.

Everything discussed in this book can be applied to licenses, though you may find there are more constraints based on how the license can be used. These limitations can give the impression that someone is constantly looking over your shoulder and in a way this is true – the IP owners will want regular progress reports to be sure that the license is being handled in the correct way. Balanced against this are the licenses that offer characters and situations which would be a dream to work with, particularly if you are able to explore them in a different way to that in the film or book on which the license is based.

One of the biggest constraints on a licensed project can be the budget. The publisher has probably paid a substantial sum to acquire the license and this money comes out of the overall budget for the game. Unless the project is a blockbuster epic, with a huge overall budget to match, the amount that remains to pay the developer to create the game can be significantly reduced from what might normally be expected. The budget for the writer is likely to be very tight with little room for manoeuvre. The writer must also take care not to create situations that will be expensive to develop. When writing for the game, part of your mind must be on the cost of everything you create – if in doubt, consult the design team.

Resources

Whatever the type and style of the license you may be working on, having access to as much resource material as possible is essential if you are to complement the style, understand the characters and match the setting. Where a game is being developed to coincide with the release of a film, this is even more important as you will have no finished product to use as reference.

Sometimes the license is simply connected with a book, like the game *And Then There Were None*, which was based on the Agatha Christie novel, *Ten Little Indians*. The only resource in this instance was the book itself, which gave a little more leeway on the visual side with the characters and setting, but imparted a strong constraint on the part of the writer who had to match a style of writing from a bygone era.

When I worked for Revolution Software, I led the writing and design of a game based on the Dreamworks film *Gold and Glory: The Road to El Dorado*. We were very fortunate to have great support and were given a lot of excellent resource material such as background paintings, character sheets, rough cut footage, the complete film script and the dialogue track. Not only did the material enable the artists to complement the film's visual style, I was able to read and re-read the script and play the dialogue over and over until the character's voices were firmly fixed in my mind. Even now, six years later, I can hear the two main characters talking to each other in my mind whenever I care to listen in.

Having such a wealth of supporting materials was of particular benefit to the team as we were under a tight deadline. For me, having established the voices in my mind I was able to write their dialogue so much more easily.

The approval process

When developing for a licensed game, getting feedback and approval for everything that is created must be managed thoroughly so it does not become a huge burden and slow down the process. The owners of the intellectual property will want to ensure that the game's developers are using the property in a manner that complements its original context, and rightly so. It is therefore in everyone's interest to establish a clearly-defined approval process with agreed turnaround times if the team hopes to be able to create a work schedule. Without this in place it is going to be extremely difficult for you to plan your writing work.

Often the process is not straightforward and a chain of approval must be followed, which will go from the writer (in our case) to the developer, then through the publisher to the owner of the intellectual property. Sometimes there may be another link in the chain between the publisher and the owner if another party is acting as some kind of agent for the property. This often means that the movement along the chain can take some time. Occasionally, the developer negotiates to go directly to the owner for approval – with other parties copied on any communication – but achieving this should not be relied upon when setting an initial feedback schedule.

Sometimes it is possible to set up a kind of rolling approval, where the writer splits up the game's story, characters or dialogue into sections and works on a different section while waiting for approval and feedback of the one just submitted. This only works, of course, when the later piece does not rely too heavily on anything that is waiting to be approved. Work closely with your development team to define a clear schedule and make sure that those involved in the approval chain buy into the process.

Another potential problem with the approval chain is the level of input from each link in the chain. With each of the parties having a vested interest in the license, a writer may have to contend with input from all of them. The danger, here is that all this feedback may muddy the waters and make it difficult to maintain a clarity in what the writer is attempting to achieve. Try to obtain some kind of agreement on the level of input from each party involved. An ideal situation would be that only the IP owner is allowed to request significant changes and the development team may go directly to the owner for feedback, but it may not be possible to negotiate this.

Adding interactivity

To make a game based upon a licensed property, interactivity must be added into a story that was originally delivered in a passive manner. It can be a real challenge to change the very nature of the medium and still retain the qualities of the original IP.

There has been some debate about the way that certain games trivialise the IP on which they are based, presumably with the approval of the property owner. For instance, if the strength of a film or TV series derives from the way the characters interact with one another, a writer can find it extremely difficult to use that strength if the gameplay, engine and tools do not allow it. Many character-driven film properties, when transferred to a game, become filled with action and violence out of proportion to that in the original. However, because the developer must have obtained approval for such an approach to the game, you should work with these constraints and build to the strengths of the original IP where you can, even if this means only doing so during cut scenes.

Maintaining the voice of the characters can be tricky when transferring from a non-interactive medium to an interactive one. In an investigative film, for instance, the main character may have very short scenes in which he gets the information he needs relatively quickly. In the game version, in order not to lead the player, the character could be asking a lot more questions which would expand the scenes. Will this fit with the character's nature and allow you to maintain the voice? Think about how you can keep the scenes tight and still allow the player to control the questioning.

If a game's story is to follow that of the original, how is it going to be transferred to an interactive medium? Scenes that simply played out in a film, say, now have to be interactive and this must be done in a way that is faithful to the original, work in the context of the game and retain the interest of the players. Many players will have already experienced the story in the film, TV series or book on which it is based, but others will be coming to the story completely fresh. You must approach the storytelling in a way that keeps the interest of both groups of people.

Because games take so much longer to play than it does to watch a film, the story that worked well in the film and felt very dynamic can feel thin in the game if it is not handled with care. Sometimes you need to create additional story material for the game to have the same strengths as the film. This may require approval above and beyond the normal process and may involve a lot of additional work and the corresponding time taken for approval. Establish these additions early, but have a contingency in case they do not meet with the owner's approval.

Massively multiplayer online games

The traditional single player video game that you buy in the shop or download to your computer is a fixed affair with a finite size, pre-set objectives and a story and gameplay that works towards a definite conclusion. The massively multiplayer online game (MMOG) is a very different and open-ended type of game in which huge numbers of players inhabit and explore a virtual world through their on-screen avatars. Gameplay may be centred on the individual or rely on players working together in cooperation. It may be highly combative or driven by quests and missions the player must fulfil to progress. Whatever the style of play, the massively multiplayer game is a beast that constantly needs feeding with new material to keep the players interested.

When the gameplay is very action orientated rather than quest and story driven, new material often simply takes the form of additional multiplayer maps and new avatar models. These may be created by the original developer and released as expansion packs, but are very often created by the players themselves, using tools released with the game, which sees them investing much more than just playing time in the experience. This additional, player-created content is often hosted by the original developer or publisher on the game's official website with the intention of creating a community that goes beyond the game itself. For many players, online gaming is not only about playing the game, but also being a part of a wider group of gamers who share a similar interest and often hang around the game's forums.

Where the game is story based or the players are able to interact in a greater variety of ways (a role playing game – MMORPG – say), additional material is generally created by the development team and the writers are an important part of that process. Although some new material will be fed into the world on an ongoing basis, many MMORPG developers release expansion packs which allow the players to explore and develop their characters in a much larger environment. The gaming community attached to many of these games can be greater, too, because as well as the online forums, the game world itself may be used to advance a sense of belonging. Some game players become almost addicted to the online world as they explore and interact with the various aspects of the game and the other players.

The writing team

Because of the scale of a story-based multiplayer project, a team of writers must be involved if the game is to have any chance of being a success. Not only must the writers work well together – matching their styles, for instance – but they must also be able to take direction from the lead writer. In turn, the lead writer must be the type of person who is able to motivate the team, work well with the design, programming and art teams, and distribute work to the other writers in a way that uses their individual talents and keeps them fresh and motivated.

In many ways, writing for an MMOG has parallels in the television industry where teams of writers will work on daily soap operas, long-running dramas and successful sitcoms. Many writers who work in these areas are often required to produce huge amounts of material each month to meet ongoing deadlines. Although some people feel that the writing teams on MMOGs should match the word output rate of their television counterparts, because of the interactive nature of what they produce, the game writer has to work with much more than just the story and dialogue. Logic that links to revealed information or conversational scenes must be sound and thoroughly tested, not only with the writer's own work, but also that of the other members of the writing and the design team.

Before the launch of an MMOG an enormous amount of material must be created so that the development team can keep ahead of the requirements to produce new material once the game goes live. This task is an immense undertaking which involves the creation of a whole world and the development of a rich and diverse back story with all the historical elements and scenarios required to flesh it out. Details will also include, but will not be restricted to, plot threads, quest details, armour, weapon and item information, character information and dialogue.

Creating the world

It is likely that during the initial stages of development the full writing team will not be involved and initially it may only involve the lead writer. Until the basic ideas for the world and the gameplay it contains have been worked out, using too many writers could be confusing and reduce the chances of creating a clear vision for the game. The lead writer – and their team when they come on board – needs to ask serious questions of the world for it to become as compelling as possible.

Does the setting lend itself to becoming a massively multiplayer game?

The world must have underlying tensions that allow for the creation of interest and drama through conflict. It must also have enough variety that players are able to choose their own path through the world and be sure to have their interest fed at all times.

The game world should be one that allows the writing team to develop a rich back story – nothing makes a player feel that they are in a compelling world more than being able to discover its complex history. More so if that history has a bearing on the current situation the world finds itself in and influences the quests the players undertake.

Hopefully the writing and design teams are able to create a world that feels and looks unique, without which it is going to be difficult for players to see what is special about the game and may well find it derivative. Even if the game is set in a traditional fantasy world, think of some way that you can put a new spin onto the world, the people, the stories and quests. What is it about your world that will not only get people playing it but make them want to continue to do so?

Thousands of people must be able to occupy the game world at the same time as one another and play in such a way that each player has the opportunity to have an exciting and worthwhile experience. The world's physical locations must be large enough to accommodate all the players without giving the feeling that it is too crowded, but not go to the other extreme and make it feel too sparse. There must be the right balance between player characters and non-player characters (NPCs), the latter being important to the fleshing out of the world and providing the player with information.

In many single-player games the player often only sees a very small part of the world the game is set in. For a multiplayer game, the world must have an epic feeling and give the impression that there is nowhere the player cannot go, within reasonable constraints. Because there are so many players inhabiting the game world at any one time, the design team must ensure that there are enough collectable items to go around and that they exist in great variety, too, for which the writers must create suitable descriptions.

As the game world is populated by a large number of player characters, there should be a broad enough variety of character choices to appeal to the numerous player tastes, which in turn helps extend the richness of the game world.

For a story- and quest-driven game, it is important that the gameplay does not simply revolve around levelling up through combat. If there are no gameplay and rewards linked to the other aspects of the world – the quests themselves, for instance – it may have a shallow feeling. For this type of gameplay the writers must work with the designers and agree the style and how it is to be presented to the player. There should be a mixture of quests that involve single players as well as teams of players.

The world should have an internal consistency which gives the impression that it is self-supporting. An apparent infrastructure in which farmers grow and supply food, merchants trade goods, blacksmiths create weapons and armour, innkeepers offer shelter and refreshment, etc., will make the environments feel as if they are a living, breathing world. If trade is an important part of the gameplay, the game world will feel more compelling if there is a sense that the trade goods originate from genuine sources and have not simply materialised out of thin air.

Although not in the domain of the writer, the locations themselves should have a consistency of graphical style so that they look to fit within the same overall world, even if they are very different in nature, such as desert and woodland. They should match the internal logic of the world so they do not look like they have simply been created for variety. Writing within this structure should complement this overall feeling in a way that makes the world feel unbreakable. Consistency across the team is vital.

The gameplay objectives, puzzles and quests should have as great a variety as the world logic will allow in both solo and team terms. Many objectives should be written and designed to encourage players to interact with other players and embark on quests as a team, for without this type of gameplay much of the real purpose of the MMOG is defeated.

Maintaining the world

Once the game is released, players need to feel that their characters exist in a living world. The writing and design teams must keep gamers fed with new material – quests, developing stories, new non-player characters to interact with, and so forth. Without this additional work, the dedicated players who spend a lot of time online will find themselves running out of things to do within a few weeks.

Although expansion packs are often released, there is usually a big gap between the original release and the first of these. The writing team is likely to split their time between ongoing content creation and expansion pack creation, although it could be that the developer requires the two areas to be kept separate and the writers involved in one may not work on the other.

However the team is structured, it is important that you do work as a team and respect the talents and abilities of the other team members. You must always be aware that the MMOG is an immense undertaking that needs an epic outlook.

Dealing with changes

Throughout the development of a game there will be many changes – big and small – which occur for all kinds of reasons. Dealing with such changes will become a regular part of the tasks you undertake and you should make allowances for this work in any schedules you are asked to produce. Without doing so, it is probable that you will end up with a serious miscalculation of the work involved, which will lead to delays as you attempt to find slots in your busy diary into which you can fit the work. You will find yourself in a situation where you desperately struggle to meet the project's deadlines.

Though it is difficult to foresee the nature and extent of any changes in advance, when estimating the writing time required you should always allow extra time for any possible changes. It is usually better to class this as a task in its own right rather than to add ten percent, say, onto every estimate you give. Simply adding time onto each task will give you additional time in the wrong places in the project.

Because an expectation of change can come across as a little negative, you should be diplomatic in your approach and call the task something like 'Editing and Polishing Based on Feedback'. For a game with a small amount of writer input, accommodating changes is probably something that you are going to be able to fit in easily, but a large project will mean a lot of change time, particularly if the finer points of gameplay or interface are still in flux. There will be games that undergo such fundamental changes of style or gameplay that whatever time you estimated will not be enough and you will have to re-negotiate for additional time to accommodate everything that is being requested.

When the changes are directly connected with the work you have done on the project you must resist the temptation to see it as personal criticism and refrain from condemning those requesting the modifications. They are professional people, too, who have to think of the game as a whole and of how the game fits into the larger context of the retail marketplace. Major alterations to your story, say, can be very frustrating and it may seem that the magic you are trying to weave has been misunderstood. Yes, you should explain and defend your work if you believe in it strongly (and if you don't then you should be questioning why this is so), but you should also be professional enough to realise when this is no longer profitable and move on. The last thing you want is for the developer to feel that they are dealing with a writer who cannot accommodate the changes they are requesting.

Many developers are creating games that push the boundaries of what is possible, technically, artistically and in gameplay terms. Story and interactive dialogue are part of that exciting mix, too, even though it seems at times they take a very subservient role to the other aspects. Because of this desire to create original games with new features, a sizeable chunk of a game's development is essentially research undertaken to find out if the new ideas will work in the context of a game. As is often the case with research, results can be unpredictable and the time involved difficult to estimate, both of which can have serious implications for other areas of the project. Even tasks that are seemingly unconnected to your writing responsibilities can have a knock-on effect.

Falling behind schedule

One of the major reasons for change is when the project falls seriously behind schedule. Most games have a fixed budget and cannot afford to miss the deadline for completion, so something must be done to avoid slippage. If caught early enough, a little additional work on the part of all team members could be enough to put the project back on track. Sometimes, though, the problems do not get resolved – or are not caught early enough – and can become compounded to a point where game features have to be removed or even whole sections of the game taken out of the design.

In a game where the story is an important part of the overall experience, losing a couple of sections of gameplay could have a major impact on the plot, particularly if part of the cutbacks is that a couple of the cut scenes must go to save time. Suddenly you are faced with a serious problem where you must either make all the plot points fit into a smaller number of game sections or to re-write the story so that you are able to reduce the number of plot nodes.

If it does not matter, from a gameplay perspective, which sections are removed, find out if it is possible to work with the designers to remove the sections that will have the least impact on the story and then build back up from there. It could be that you have sections that are only relevant to sub-plots and it may be possible to remove these without a serious overall impact.

Cuts caused by time and budgetary restrictions are going to make it difficult to recreate the original scenes in a new area and the writer must find new ways to present the information or be forced to remove it altogether and simplify the plot. Some plot information from the sections that are cut can be incorporated into other scenes, though care has to be taken that the scene structure is not undermined through doing so. Sometimes the writer is forced to put the information into a diary, letter or other such device where it can be read as text. In my view, this is rarely a satisfying way to present plot information and should only be used if it is the only solution available within the new restrictions.

The principle of 'show, don't tell' should be extended to 'show, don't tell, don't use diaries'.

Market perceptions

The games industry changes rapidly and sometimes there is a great temptation for publishers and developers to react too strongly to a perceived market shift. If the level of conviction on a project is not particularly high then developers could be looking for a winning formula that can be applied to the game.

The gaming press is constantly filled with the next big thing where journalists become excited by, hopefully, outstanding games to the point where perfectly good games are ignored. While there is nothing wrong with presenting news and coverage of games, there can be the temptation to compare other developers' games with your own. Familiarity with your project can be a real danger, making the screens and gameplay everyone has been staring at for months seem a little stale in comparison to the screenshots you have just seen in a magazine and the article you have read on the gameplay.

The temptation to tweak and change in response to the perception of what is the current hot game must be resisted or it will add a great deal of time to the schedule. It is important that the team holds true to the original vision, including the writer's part in that vision.

However, there are times when the publisher is willing to put in additional funding and push back deadlines so that certain aspects of the game can be changed or adapted. In a development cycle which could last two years or more, there is always the chance that the game will be perceived as being too 'last year' even if it was designed to be cutting edge when the project started. Publishers are ever mindful of the rapidly-changing nature of the marketplace and even the most experienced can have a difficult time forecasting two years ahead.

When the request for changes comes from a need to fit into the market, you must be aware of recent games and those on the horizon. Understand their flavour, their gameplay, what they are doing with story and dialogue, particularly those that are close in style to the game you are working on. At the same time, though, try to change and adapt the story and characters so that they offer something original and fresh.

Character or story changes

You get a phone call asking you to come in to the developer's offices for a meeting to discuss the game. On arrival you find that story discussions have been taking place of which you were unaware and now the main character is no longer a teenage male but a female in her mid twenties. After you pick your jaw off the floor you quell your anger and ask what prompted such a major change.

There could be a number of reasons why this has happened, none of which will probably feel satisfactory. Any major changes of this nature are going to have a big impact on the story and, depending on how important that story is, on the game as a whole. Be sure that the developer understands the amount of work involved as a consequence of the major changes being requested. Even if you have allowed time in your schedule for work of this nature, it is unlikely that you have been expecting it to be on such a scale.

One thing you should not do (unless the developer gives you no option to do otherwise) is to cram the new ideas and changes into the original structure. A major change to the main character will probably mean that you need to go back to your original plotting notes and build up again from there.

At each step of incorporating the changes you should think through the implications. A seemingly straightforward request could have a big effect if it means changes in each section of the game, particularly if the game is non-linear. Understanding the structure and flow of the game along with its story and characters is even more important when it comes to dealing with the changes requested.

Bug testing and fixing

All games need to be thoroughly tested before release – nothing spoils a player's enjoyment more than a game that pulls them out of their immersion because of graphical oddities, technical glitches and gameplay oversights. The story and dialogue in a game can become potential problems if not properly tested and the bugs eliminated. Hopefully, as the writer on the team, you have been involved in the ongoing process and have been playing new builds of the game at regular intervals and have caught most of any dialogue or story problems that could otherwise end up as bugs.

For those who have never been involved in the testing of a game, good games testers will try playing through the game in ways which specifically look at possible ways to 'break' it. They will play small sections over and over, trying different combinations of actions and change the order each time they play through. This process will test the logic structure and flow of the story and dialogue very thoroughly and any weaknesses will be thrown up as bugs.

Writing bugs can vary from simple spelling or grammatical mistakes to whole scenes that do not make sense because of the order of play. Spelling mistakes are easily corrected, although if you have created a number of original words, particularly names in a fantasy epic, say, you need to be sure that there is a consistency of spelling throughout the game. If the game has a team of writers, it is important that they all spell 'Zorak' in the same way or players could think that there are multiple characters with similar names.

Dialogue, particularly when tested before the voices are recorded and only seen as on-screen text, can be a source of bugs which the tester thinks are grammatical errors. If a character has a stylised way of speaking – and many people do – you may be ignoring the rules of grammar to write their dialogue in a convincing way. If bugs are thrown up in this situation, you probably need to write a note explaining this and have the bug closed.

Sometimes dialogue bugs can be attributed to the variables controlling the scenes. If the player constantly triggers the same snippet of dialogue from a character when it is clear that the conversation should be moving on, then it is probably because a variable has not been set in the script. You may never see such a bug in your report as they are usually handled by the designers, but if you do then it will be something you should discuss with the designer or implementer working on that part of the game.

Where a bug throws up the need to change a scene because it does not make sense, or fails to convey the right meaning when played in a different order, the changes you undertake must be handled with care and the consequences worked out completely, ensuring that other scenes are not affected adversely as a result of the change. It is very possible when fixing this type of bug that you actually create more bugs as a consequence.

Where your change requires alterations to the logic structure of scenes or additional small dialogue interchanges, be sure to work with the designers involved to ensure that they are aware of all of your changes. This also applies if you are part of a writing team if the modifications have a wider ranging impact and stretch beyond the section on which you are working. You do not want to create a situation where your bug fixes throw up bugs in another person's section.

Most development companies have software that tracks the progress of bugs and when fixed will tell the person who reported the bug that this is so. They will then be able to test the fix in order to check it off the list. Not only must you fix the bugs, you must also report it as fixed. The testing period can be a very tense time and fixes must be done as quickly as you can so that all runs as smoothly as possible.

Recording the voices

At last you have arrived at the point where there are no more changes that affect the dialogue – the bugs have been fixed and everyone is happy with what you have created. You breathe a sigh of relief as the scripts are locked and preparation for recording can begin. That could be the end of your involvement in the project – job done, as it were – but it may well happen that you become involved in supporting the recording of the voices, too.

Ideally – and particularly for a game with a lot of dialogue – a writer should be on hand during the recording because there are frequent situations which require the writer's skills and knowledge. Even well-written scripts cannot always put the scenes into a proper context for actors who have had no contact with the game before entering the studio. The writer should be available to answer their questions, offer support and resolve any issues that may arise.

Script preparation

You have worked on the dialogue scripts so intently that you know them inside out. You have gone over and over them, testing and re-writing, developing the characters until the game is a vibrant work filled with your dynamic dialogue. So who better to prepare the scripts for the actors?

How you do this will depend on the way the lines are to be recorded. Many game projects simply record one character at a time, booking the actor as required during the recording period. For a more dynamic recording session, though, actors may be required to be in the studio together – an ensemble recording – which gives them the opportunity to feed off each other and fill the scenes with a vibrant, more naturalistic flow.

Whichever recording method is being used, you must always have a master document that contains every scene in the game, including all the one-liners and brief descriptive comments. You will need at least three versions of this – one for yourself, one for the voice director and one for the sound engineer who will later use it to check and create all of the individual line samples. Master documents are useful for ticking off each line as it is recorded, making notes of changes and keeping track of which of the recorded takes is the right one to use. For some games, the master document can run to hundreds of pages, so it is useful to break it down into manageable chunks organised by recording session or by actor, matching the organisation of the studio sessions. As each portion is completed it should be quickly double-checked and then put to one side in preparation for the next one.

The formatting of the scripts may be a little different from what the actors are generally used to because it is likely that each line in the scripts will be tagged with a number in the tools. This is usually a process that takes place once all the changes have been made and the scripts are locked. Identifying the lines by number is necessary because each of them will be substituted when the game is translated. If the tools have been set up with recording and translating in mind (and it is always worth checking this when you start on the project) they could be exported into a format that could look something like the following:

Scene – Edwards talks to Wilks				
	[20201]	Edwards:	Hey, Wilks.	*//Greeting used every time*
First Time				
	[20202]	Wilks:	What's up?	
	[20203]	Edwards:	I heard that you witnessed the shooting.	
	[20204]	Wilks:	That so?	*//Nervous, but puts on brave face*
	[20205]	Edwards:	Just tell me what happened!	
	[20206]	Wilks:	Get lost! I didn't see nothing!	
				//Edwards looks angry – he should have handled it better
Other times (not talked to Wilks about Johnny)				
	[20207]	Wilks:	I got nothing to say.	
Edwards has talked to Johnny				
	[20208]	Edwards:	Your friend Johnny saw you with the body.	

	[20209]	Wilks:	That junkie ain't fingering me!	
	[20210]	Edwards:	It's not looking good, man.	
				//Wilks thinks, weighing his options
	[20211]	Wilks:	Look, all I saw was a guy in a leather jacket running away. The woman was already dead.	*//Leather jacket is key info*
	[20212]	Edwards:	Thanks.	
Repeated response line when all information is obtained				
	[20213]	Wilks:	Get lost, will you?	

Because it is easier to keep track of the lines, numbers, comments and translations in a spreadsheet or database table, this is how it is likely to be exported, but because the above is not very readable, particularly for actors who need to read their lines easily in the studio, a little careful formatting of the columns and hiding the grid lines can give you something that is much closer to a more traditional script:

Scene – Edwards talks to Wilks

[20201]	**Edwards:**	Hey, Wilks.	*//Greeting used every time*

First Time

[20202]	**Wilks:**	What's up?	
[20203]	**Edwards:**	I heard that you witnessed the shooting.	
[20204]	**Wilks:**	That so?	*//Nervous, but puts on brave face*
[20205]	**Edwards:**	Just tell me what happened!	
[20206]	**Wilks:**	Get lost! I didn't see nothing!	
			//Edwards looks *angry – he should have handled it better*

Other times (not talked to Wilks about Johnny)

[20207]	**Wilks:**	I got nothing to say.	

Edwards has talked to Johnny

[20208]	**Edwards:**	Your friend Johnny saw you with the body.	
[20209]	**Wilks:**	That junkie ain't fingering me!	
[20210]	**Edwards:**	It's not looking good, man.	
			//Wilks thinks, weighing his options
[20211]	**Wilks:**	Look, all I saw was a guy in a leather jacket running away. The woman was already dead.	*//Leather jacket is key info*
[20212]	**Edwards:**	Thanks.	

Repeated response line when all information is obtained

[20213]	**Wilks:**	Get lost, will you?	

Obviously, the exact layout will depend on the format of the game's scripts and how they are exported. How well you are able to adjust the formatting to something that is reader-friendly will depend on your knowledge of spreadsheets, so gaining familiarity with them could be very valuable to you. If the final scripts were exported into a Word document, say, it would take much more work to format them into something the actors could work with. With a spreadsheet it is possible to format each column quite quickly and the whole document is done in a very short time. If the scripts are broken up into a number of different files or different pages within the spreadsheet, it may be worth learning how to record a macro for the formatting, which will make the whole process even easier.

If the game's export tools are really good, it could be that they can organise the scripts into game sections for you. Even better would be if some kind of filtering was applied so that the scripts can be exported by character name. This way you can organise the actor schedule according to the character scripts.

If you are recording the actors one at a time, each actor will need a copy of all the scenes in which their character appears. Ideally, the game's programmers can help you with this filtering, otherwise it is probably down to you to organise this manually in some way. Once all your scripts are prepared this way, the schedule can be organised with the voice director and studio, then copies of the scripts should be sent out to the actors for familiarisation. When you arrive at the studio, do not assume that the actor will remember to bring along his or her script – ensure you have enough spare copies for all the actors to use in the studio.

When recording with an ensemble cast, it is a little more difficult to organise the schedule and minimise the time that actors are waiting around. Filtering the exported scripts by character/actor is even more important here as it can help you identify actors who overlap with one another. You should always identify your important characters and actors and filter for their scenes first, followed by any characters that overlap with them.

It could be that Character A is not in any scenes with Character B, but both are in scenes with Character C. This could mean that the actor playing Character A does a morning session and the actor playing Character B could come in for the afternoon session. The actor playing Character C would be in all day. However, if the number of lines is not very great for any of them, it may well be that all three are finished within an hour. Sometimes it is best to have some of the other actors doubling up on the minor characters to maximise efficiency, but where possible try to keep them to different areas of the game.

If a script is very complex it can mean that you get other actors also overlapping with the three above and planning is an essential part of this. I have had days in the studio where there were eight actors rotating with each other to cover all their overlapping scenes. The session passed incredibly smoothly because of good preparation and the professionalism of the actors involved.

The voice director and actors

The advantage of a development studio hiring a dedicated voice director or using a recording studio that offers this facility is one of quality. Good directors will be able to get that little bit more out of the actors, but only if they have prepared well and appreciate what you want from the recording. The voice director and studio must understand the requirements of interactive dialogue scripts.

It is important that you work with the director when preparing the scripts and the schedule so that he is fully aware of how you are breaking down the recording sessions. It may well be that he has his own ideas for organisation that needs to be taken into account, particularly if the actors involved have limited availability.

A good voice director or studio will have the right actor contacts to suit the type of work involved in recording for a game. It is important that the actors understand that games are an important, developing medium and give as good a performance as if they were in a stage, film or television production. I have known quality actors who have never done game voices before, but once they understood the richness and variety they thoroughly enjoyed the process and really entered into the spirit of it and enhanced their roles.

The voice director should encourage the interest in the game's scenes. How the characters interact with one another and progress through the game should be emphasised where appropriate.

In the studio

If you are in the studio during recording, there are a number of things you should be able to do.

The first is to follow the scripts on your master copy and be sure that they get every line down. It is easy for an actor to misplace pages or to turn over more than they realise because he or she is concentrating on the performance. Some lines are almost duplicated with subtle variations that actors may miss, thinking they have just recorded them. Notes should be taken at all times where relevant about alterations, odd pronunciation, and so forth.

You should support the voice director by listening to the delivery of the line and be mindful of the tone in case there are times when the context of the scene has been misunderstood. Be very diplomatic – the director should be running the show in the studio and interruptions should be kept to a minimum. However, the director is unlikely to know the game's story and dialogue like you and will probably appreciate any enrichment that you can give to the actors, but always ensure that instructions to the actors is passed through the director.

You must be able to resolve problems on the spot. Sometimes, for example, a line looks fine on paper but when the actors speak it out loud it does not sit well. Re-writing the line, often with the aid of the actor and director, is something you must be prepared for. Because good actors really get into their roles, I have actually known them to spot inconsistencies that testing has missed. Fortunately this has only happened a few times and all have been minor and easily fixed, but it shows how you must be aware of the whole preparation and recording process.

The studio engineer will ensure that each line is recording at the correct level and in a proper fashion. It is important that everyone involved understands that each line must be recorded without overlap because after recording is complete, each one is saved as an individual sample. Sometimes, when an ensemble cast are moving along rapidly with a scene, they can forget this and the engineer will usually catch instances where overlap occurs. If you feel that the engineer has missed an overlap, ask if you heard it correctly. The engineer can usually replay the lines involved very quickly and the interruption is usually minimal, but it is always better to be sure than to risk the lines being a problem at the sampling stage.

Double checking

Recording the voices is an important part of the game's development and you will probably only have one shot to get it right. The expense of pulling people back in for re-takes is not something the developer would like to have to pay for. Preparation is an important part of this, but you should also double check everything during your time in the studio.

Before you let each actor finish, quickly go through their scripts to be sure that you got all their lines. Check that the engineer has no problems and that the director is happy to let the actor go. Occasionally the director may want to re-take certain scenes later in a session if there is time. If this happens, remind them of this and ask if they still want to do it.

When you have double checked a session and everything is in order, put those session scripts into a 'completed scripts' folder on one side. This means they are available for reference, but reduces the clutter as you work through the recording.

After the recording

Once the recording is completed the engineer will cut up the samples and deliver them to the developer. It may be that your involvement is now over, though it could be that you are asked to review the samples before they are placed into the game.

The samples should all have been checked for consistency of sound level and adjusted where necessary, but sometimes this is missed and you need to check. There are occasions where multiple takes have been given because the engineer was unsure which to go for. This is where the notes taken during recording will be invaluable and help you decide which of the takes to use.

Once the samples are checked, the development team will place them into the game and you will at last get to hear them in a proper context. At last you can enjoy the fruits of your labour as the game truly comes to life.

Localisation

Localisation is the term used when versions of the game are created for territories that have different language needs. The word 'territories' is used instead of 'country' because a specific language version may be used in more than one country. An English language version, for example, would also be valid for Australia, New Zealand, Canada and the United States. However, because of differences in television hardware and spelling discrepancies – between the US and the UK, for example – separate versions often have to be created to match the target requirements. Though these are not full-blown localisations, as far as building a release version of the game they are often treated as such, organisationally.

US versions

When developing console versions of a game, the US market has different technical requirements to European territories because their TV input is NTSC instead of PAL. Of course, the reverse is true and US developers must also deal with the different technical challenges when creating versions of their game for non-NTSC territories. When supporting multiple console platforms, a version for each of them must be handled on an individual basis and tested completely in its own right. Sometimes technical issues can throw up more than just superficial differences and sometime bugs occur that never surfaced in the original swathe of versions.

In terms of written content, creating a version for the US market is not just about finding the words that are spelled differently – anyone can replace 'colour' with 'color' – but also identifying the differences in terminology. 'Petrol' becomes 'gas' and 'spanner' becomes 'wrench' are common examples of the variations that occur. Although many of these are known to us through a worldwide proliferation of US films and television programmes and the fact that many of us regularly chat with others across the world, it is still unlikely that we will be completely familiar with all terminology differences. However, a swift search on the internet can usually turn up the information you require.

For instance, in Britain, 'bottom drawer' is a mostly outdated term, traditionally referring to the place where a young woman will place the items that she is saving towards her eventual matrimony. I wanted to know if this phrase was used in the US and a search on the internet turned up the term 'hope chest' as an equivalent.

Even if you discover all of your parallel terminology, it is not necessarily desirable to use it. A character in the game who is English, for instance, would not use American expressions and vice versa. I once worked as script editor on a game based on an Agatha Christie novel because the US company developing the game wanted to be sure that the English dialogue sounded British and not American.

Occasionally, the differences in vocabulary bring about a smile. In Britain our kitchen sinks have taps attached to them, while those in the United States have faucets. However, apparently, the people of both countries refer to the liquid that comes out of them as tap water. This also shows that care must be taken not to make the wrong assumptions about the words used.

Full translations

French, German, Italian and Spanish are common languages for translation in the European game market. Other languages are decided on factors which set the number of possible sales against the cost of translating, recording and testing that version.

Regardless of the number of translations taking place, the scripts should be presented in a single format, which usually means taking the spreadsheet files that you used for the recording scripts and changing them a little to accommodate an additional column for the translated lines. The column entitled 'Translation' would be changed to whichever language the translator was working in.

English **Translation**

Scene – Edwards talks to Wilks

	[20201]	**Edwards:**	Hey, Wilks.	*//Greeting used every time*	

First Time

	[20202]	Wilks:	What's up?		
	[20203]	Edwards:	I heard that you witnessed the shooting.		

			English		**Translation**
	[20204]	Wilks:	That so?	*//Nervous, but puts on brave face*	
	[20205]	Edwards:	Just tell me what happened!		
	[20206]	Wilks:	Get lost! I didn't see nothing!	**Meaning: I didn't see anything.**	
				//Edwards looks angry – he should have handled it better	

Using a spreadsheet for translation ensures that no lines are missed by the translators and also means that when the translated files return the programmer responsible for the localisation is able to import the new versions of the lines into the game more easily than having to copy and paste each one.

Before these scripts are completed, however, it is a good idea to go through them and look for possible ways that misunderstandings could occur. In the table above, line 20206 is not grammatically correct, but it is the way that some people speak. A good translator will probably know this, but as it does not hurt to make sure I entered the additional comment (in bold).

Not only can dialogue styles cause problems, but also the context of a line. There may be, for instance, a line in which the player character comments to himself and says, 'I'm not sticking my head in there!' The translator may not know the context and will not understand what 'there' is referring to. Because gender is attached to nouns in many non-English languages, to make a correct interpretation, the translator must know whether the object in question is a hole in the wall, a toilet, a pan of boiling oil or any other of a million things in the game world. Although you have probably covered most of these instances with comments designed to aid the actors, be sure that your notes cover the context in any other event that could be seen as unclear.

Once the scripts have been completed in this manner, a copy of them can then be sent to each of the translators and the person overseeing the localisation will ensure that it all comes together correctly.

In recent years there has been an increase in localisation studios that specialise in game translation and often include the recording of the translated dialogue as part of the services they offer. The advantage of this from the game developer's point of view is that the whole process becomes more centralised. If there are any questions about the scripts, they are usually coordinated by the localisation studio's representative and can quickly be passed onto the translators of all versions.

You should be prepared to answer the questions that come from the translators, either through the representative, if using a studio, or from each of them individually if that is how the translation work is being carried out. Even though you may have commented the scripts thoroughly, there will always be a few lines or scenes that need further clarification and your understanding of them is better than anyone's. It is vital that you answer these questions as swiftly as possible – the localisation of the game will be on a tight schedule and it is important that none of the deadlines are missed.

Script editing

There will be times that you may be asked to work on scripts that have been translated from another language into your own. Although translators in general do an excellent job, in a game that relies heavily on characterisation and the interaction between characters, if the style is not right for the language the dialogue may come across as a little dry. This can be disastrous for humorous games that not only rely on jokes and one-liners, but also on the tone and presentation of the dialogue in general.

When editing a translated set of game scripts you need to ensure that you understand what the original creators' intentions were. As you read through the whole of the dialogue you must try to see through the translation to what they were trying to say in each of the scenes. If possible, you should try to get hold of the original game or voice samples so that you can hear the tone of the dialogue, even if you cannot understand the language. Even more advantageous would be to get hold of an original build of the game and play it through so that you can see each of the scenes as originally intended.

Once again, much of this relies on understanding the context and tone. A simple line like, 'You've got to get out of town' may be a threat, it could be advice, or perhaps the character is pleading with the person they are talking to. Usually, the way the scene plays out will establish the context, but if the lines spoken are in subtle conflict with the real meaning of the scene you may need to be clear what that meaning is meant to be.

Where a scene is a humorous one, the humour can often be lost because the rhythm of the exchange between the characters is not quite right. Although it is rarely possible to change the number of lines in a scene – because the game's implementation relies on the scene's structure remaining the same – you can still improve the rhythm by re-writing the individual lines. Sometimes you will even need to re-write the whole exchange to make it work, but when doing this, the characters' lines must remain in the same order and are usually required to be of the same length as the original. Be wary, though, that you do not lose information that is important to the plot or to the success of completing the game.

Sometimes, if a joke is translated literally it makes no sense or just fails to be funny and the joke must be re-written or another substituted. If the joke refers to an object in the location, then however you re-write it, it must still be about that object. It could be that the joke has been triggered by the player interacting with that object in some way. The same applies to any dialogue which references specific objects – any editing must retain the reference.

A note on timing

While there have been many poorly-translated games in the past, now that there are dedicated game translators this is something that is becoming less common and the overall standard is very good. However, there are still games that can give the impression they are poorly translated or badly voiced.

I recently played a couple of best-selling games that had been translated into English from the original Japanese. Although the games were excellent to play, the dialogue scenes came over very badly. When I studied them closely I realised that it had nothing to do with the translation directly or the voice acting, but the timing of the exchanges. It was clear that the timing of the scene still matched that of the original version and had not taken into account the length of the English dialogue lines.

Some games have their dialogue engines constructed so that as soon as one line sample finishes the next one in the scene is triggered. So if the length of the lines is different in each of the localised versions this does not matter as the engine accommodates this. Increasingly, though, cinematography, facial expression and body language are being used in games to give the player a richer experience and allows the writer more scope to add emotion and subtlety to the scenes. In most examples where this occurs, the timing is something that is fixed throughout the scene, along with the playing of each dialogue line.

In the games I played, the translator and/or script editor probably thought that the scenes would be improved if the dialogue was tightly written. Unfortunately, because the timing did not accommodate this approach to the dialogue, the actual effect created was the opposite. The scenes were filled with long pauses while the original timing waited to trigger the next line of dialogue, which gave each scene a very non-dynamic feel.

Clearly, it is important to discover if the game engine can adapt the scenes to the timing of the lines or if the timing is fixed and each of the localised versions must have line lengths which match the original. This is information that is important to pass onto the translators of any game that you have worked on as a writer, but also will affect how you approach the script editing of a game that has been translated into your language.

Maximising the quality is not just about writing, translating or editing to the best of your ability, but also about matching the limitations of the game engine.

Technical writing

Some years ago, those involved in writing for games came from a variety of backgrounds and often ones that had a more technical bent. The early history of game development was filled with programmers who created games completely on their own, originating all the graphics and writing the story and in-game text if the game required it. Many of these people went on to head successful development studios, specialised in areas of programming that suited their particular skills or concentrated their career development on the game design side of things. Some are still involved in aspects of game development that involve writing, though not always directly involved with the story or dialogue.

Writers with a technical background or a strong understanding of the technical side of game development can be very valuable. This knowledge may have come from extensive experience or simply a strong interest in technical advances and progression and how they relate to game production, but however the know-how was obtained the technical writer could be in a position to help a studio meet its project deadlines during the pre-production phase.

The technical lead on a project is often swamped by a number of vitally important tasks – leading and supervising the other programmers, researching the technology for the new game, developing schedules, helping to create the prototype demo and writing the technical documents. The writing of the documentation can be very time consuming, so if this is placed in the hands of someone else, the technical lead is then free to use their time and skills to better advantage.

The technical design review

Technology in games, like other areas, is highly specialised, particularly where innovation is concerned. Many big-budget or leading-edge games succeed or fail because of how they use advances in technology to deliver exciting, fun gameplay. A developer must convince a publisher that their understanding of this technology is strong if they are to obtain the funding the project needs to carry it through development to completion. With games becoming increasingly expensive to make, publishers are becoming ever more cautious about the projects upon which they lavish their money. If a large part of the development budget is earmarked for technical innovation, it is important that the publisher understands what that money will be spent on and what will be delivered at the end of development.

The technical design review (TDR) is a collection of documents that outline all the technology – hardware and software – that will go into making the game the best ever. It should cover, where appropriate, the player interface, the use of simulated physics, the audio system specifications, graphics rendering techniques, the use and incorporation of middleware, outlines of the various target platform differences and how they will be resolved and many other details specific to the project.

Because a lot of time, thought and expertise goes into developing the ideas and systems involved, the documentation can run into hundreds of pages – something that will take the technical lead a great deal of time to pull together. A writer with the right background and understanding will be able to help take much of this burden, using notes, conversations and reference to pull it together.

The TDR serves two purposes. Internally, it serves as a description of all the technical tasks that will be required during the development of the game and gives the technical team, and others, the vision of what the game should deliver. Externally, when combined with the other game documents, it is a sales brochure designed to convince a publisher to invest in the project.

To have a chance of persuading the publisher that the technical team is competent, the TDR must be written in clearly, explaining every aspect of the game's technical development. If there is even the slightest chance that an explanation or detail may be misunderstood, then it must be resolved so there is no ambiguity.

It is very easy to fall into the trap of assuming that the reader of the TDR will have the same specific knowledge as the technical team on the project. This might well be the case, but with the rapidity of software and hardware advances it is always worth explaining the concepts behind the technology, perhaps as a side bar, so that the publisher understands that the tech team knows what it is doing. This is particularly important where the technology is pushing at the boundaries in some way or that the software being developed is proprietary.

Often the technical development is not just about creating logic engines, rendering engines and the like, but also about developing the tools that enable the implementation team to put the game together through the use of level editors and scripting engines. Where the development of such tools is vital to the project, it must also be covered by the budget and so becomes part of the TDR.

Internally, the advantage of a clearly-written TDR is that the project manager is able to read and grasp it so that they can create a comprehensive schedule, in conjunction with the technical lead. Because a schedule is often an important part of the game's proposal and will form the basis for milestones and the payments that are made upon each one's completion, having a complete specification of the work involved will give the publisher the confidence that the studio can develop an exciting game and do it on time and within budget.

The technical summary

While the TDR is vitally important, some of the people involved in the decision-making process, within the publisher's organisation, are not likely to be so technically minded. Even if they are, they could simply wish for a summary document that enables them to understand quickly the technology the studio is proposing to develop and use.

The technical summary should be less concerned with how everything is going to be done – which is the purpose of the TDR – and more with why the work is necessary to give the game an innovative look and feel or how it will make the gameplay more immersive and addictive. Of course, some explanation is necessary on each point, but should be limited to a paragraph at most and take a very high level stance, being explained in broad terms without going into fine technical detail.

The technical summary should be written in a dynamic manner that encourages the reader to become excited about the technical proposals. It should give the impression that the development studio has complete confidence in their abilities and in how the proposed technology will make the game an exciting one to play and a very marketable product.

It is important that the publisher feels that their investment is going to be worthwhile. With a lot of money at stake – millions of pounds in many cases – publishers have to be confident that they will sell enough copies to cover their investment and give them profit on top of that.

Strategy guides and manuals

Strategy guides and manuals have a little in common in the way that they are designed to help the player understand how the game works and what they are supposed to do. Explaining such things as installation, the control layout, background details, weapon use, inventory management, etc., may not be necessary for every player, but for those who need the help it is available. While the manual comes free with the game and is generally a small guide that offers just enough help to get the player into the game, a strategy guide covers the whole of the game from start to finish, showing the player how to play successfully every section of the game through the use of maps, screen-shots and clearly-written instructions, which cover descriptions of traps to details of secret areas and the key combinations for special moves.

Although writing these books is not, strictly speaking, writing for the games themselves, for the writer who is also an avid gamer, writing strategy guides and manuals can be an exciting way of combining writing their skills with extensive game playing.

The manual

A game's manual, which is included with the game, must cater for both the experienced and novice gamer. It should be laid out in a way that introduces the newcomer to the whole process of installing and playing the game, the use of the control interface, saving and loading, how the player character fits within the world, the player's objectives and so forth.

The experienced gamer will rarely even look at the manual, unless it is to consult diagrams on the layout of the controls or features that are not standard on other games. They will have an understanding of how games work that is built up from a lot of time playing a variety of titles and so they often only look for the details that make a game different.

A manual must always be written with the novice in mind, offering instructions with a clarity that should never become patronising. Instructions should be brief, but clear, using icons and other graphics lifted directly from the game's resources to aid those instructions.

Some games have very complex systems as part of the game engine and the player may be expected to manage character stats, manipulate and manage large inventories or organise the building of strategic teams with all the attendant attributes. A series of screenshots from the relevant parts of the game will be used to illustrate these features, but the text description will take them through the in-game process and how it works.

Some manuals, to help the player get a proper feel for the game as quickly as possible, include a small walkthrough of the first ten or fifteen minutes of gameplay. It is important that this is written in a way that not only ensures that the player understands why they are doing the actions described, but also in a way which reflects the style of the game.

Do not forget that games are created for a number of target platforms – computers, home consoles and hand-held consoles. Each will have their own manual requirements, particularly those connected with discussions of interface controls, saving and loading, and menu navigation. Unless any descriptions or other details need to reference the platform specific information, a very generic approach should be used when writing so that it can be applied to all versions of the manual and keep any specific editing down to a minimum. So, for example, avoid phrases like 'click on the icon' as this suggests a PC-orientated mechanic. The phrase could become 'select the icon'.

The strategy guide

A strategy guide is, usually, a large format book that is created and published by a company other than the game developer or publisher. Although unofficial guides have been published in the past, this is very rare nowadays and most guides are official ones. Usually the guide publisher pays the game publisher a fee for the license to publish the guide.

In some ways, the term 'strategy guide' is a bit of a misnomer, suggesting that the book will help the player develop a strategy which will allow them to complete the game. However, many games simply do not have that variety of gameplay and mostly have only one way of overcoming each of the gameplay obstacles. The guide then becomes a way of helping the players to get past the obstacles they become stuck on or shows them areas they may not have discovered on their own. It effectively becomes a way of making the game easier for the player by walking them through it.

Evidence suggests that if a game has an accompanying strategy guide it sells more copies. This is clearly advantageous to both the publisher of the game and of the guide. The guide must be published at the same time as the game for this to be most effective. This means that the writer of the strategy guide must have access to the game well before it is released to be able to complete the work.

Because this access will usually take place during a period when the game is going through a final testing and polishing phase, any last minute changes in the game could affect the content of the strategy guide. However, generally speaking, the last couple of months before gold master are usually spent making minor adjustments and fixing bugs, so any changes that affect the guide should be quite minimal, all being well.

Changes, no matter how small, could still affect the guide, so it is important that you come to an agreement with the developer that you will be kept up to date with any changes that take place. The main problem here is that the developer will be extremely busy during this period and you may be competing for even the smallest amount of their time with many other things of much higher priority. Even the basic information that you require may be difficult to obtain at short notice, so be sure to plan your requests with as much time leeway as possible. It is always best if you can have one point of contact, preferably the in-house producer or project manager as this person should be the one who can most easily pull together any resources that you will need without affecting the programmers and artists who will all be extremely busy. Be sure that you obtain information on all the secret or hidden areas that may be a part of the game.

The way you approach the writing of the guide will depend on the style of the game, its length and its complexity. You should look at other published guides that cover games of a similar genre to get an idea of what is expected and how it should be laid out. This will also probably inspire you to think of how you may be able to put your own stamp on it to create a guide that is both informative and a pleasure to read.

Most guides are great to look at with plenty of graphics taken from the game and the book is laid out in a well designed manner, so your writing must compliment this. Even with the usual large format of the pages there is no point in writing five hundred words for each page or it will leave no room for the graphics. Your writing must be tight and to the point – the buyer of the guide wants it to help them work their way through the game and too much text will be off-putting and keep them from playing the game for too long.

The advantage of keeping the writing brief is that the whole book should not take too long to write, which is often a necessity. A writer is often given only one or two weeks to complete the book, so a balance between clarity, completeness and detail should be aimed for.

The guide must take the player through the game from start to finish. How much detail is to be inserted along the way will depend on the amount of time you have at your disposal and how much support you get from the game developer and publisher. Plan your work carefully as it is vitally important that you deliver your manuscript on time.

Writing strategy guides is not for everyone and the writer who goes into it must really enjoy the task and be prepared for a lot of hard work over a short time period.

3 You as a games writer

Chasing the work

If this book has succeeded in its intention, you should now be in the position of understanding how a skilful writer fits within the game development process. When you combine this knowledge with your skills as a writer you should be in a strong position to begin the process of looking for writing work in game development.

Because writing for video games is still in its infancy, it can be difficult to find developers who fully understand the value of an experienced writer and what that person can add to their project. Admittedly, the situation is improving – the field of interactive writing is progressing all the time with standards rising and the end quality improving, but we are a long way from the position where the use of a dedicated writer is a standard part of game development.

Identifying the client

Not only will you have to find developers with new projects starting up, you will have to be sure that those projects are ones that require the skills of a writer. Games like *Sonic the Hedgehog* or *Zoo Keeper* have little text or dialogue requirements, so be sure that you have done your research properly and that the studios you approach create games with writing needs. You may well need to think ahead – a company that has a project nearing completion is probably already in the planning stages of the next. By keeping abreast of the latest game development news and release schedules you will be able to identify companies that you can target. One excellent way is to subscribe to Games Press (www.gamespress.com) or by regularly visiting gaming sites like Eurogamer (www.eurogamer.net), Gamespot (uk.gamespot.com/news/index.html), Games Industry (www.gamesindustry.biz/index.php), Women Gamers (www.womengamers.com), Gamasutra (www.gamasutra.com), and a whole host of other sites. The two magazines in the UK, Develop (www.developmag.com) and MCV (www.intentmedia.co.uk/publication.php?id=11523) are also excellent.

Once you have found such potential opportunities, there will be occasions where you will not only have to sell your own skills and experience, but also convince the people involved that their project will benefit greatly from what you can bring to it. Many developers see writing as a much less important aspect compared to design, programming and art, but if their game is one that uses story and dialogue as part of its gameplay you should attempt to convince them that what you offer will add enormously to the quality of the game, improve the review scores and help the overall sales.

Of course, there are plenty of developers who value good writing but may feel that they have this covered by their in-house staff. This may well be the case, but if you approach them diplomatically you may convince them that your additional experience will add further richness to their game. Often in-house writers have a dual role and do this work alongside their main role as game designers. While this does not preclude them from being a good and experienced writer, their design duties may seriously reduce the amount of time they are able to spend on the writing which could suffer as a result. One possible approach when in discussion with a developer is how a dedicated writer has none of the other, highly important, tasks that could act as a possible distraction from producing quality writing.

One of the hardest parts of obtaining writing work is getting yourself in front of the developer. Be very wary of how you approach developers and always do your research so that you understand the type of games the company creates and you have contact names to approach directly.

Sending cold calling e-mails can be fraught with problems because without the right approach it may be seen as spam and you have lost your opportunity. Not only will you not get any work, you may not even get a reply, which is not necessarily rudeness on their part but a failure on yours. If you approach the developer by e-mail, make sure you word it so that it does not read like a form letter – not only would this show you in a poor light, it would also be rather insulting to the person receiving it as most people have the ability to spot a form letter.

Cold calling is never going to be a good way to find work as most people are very unreceptive towards it. You therefore need alternative ways to look for the writing work: consider becoming involved in the gaming community to some degree or other.

Joining the International Game Developers Association (IGDA) can be very useful, particularly if they have a chapter which is local to you. The IGDA is an excellent resource for all kinds of game development information, and also has specific information for writers in the form of the Game Writers Special Interest Group, which gives the opportunity for game writers to chat with each other online or to meet up at organised events during game conferences and shows.

The IGDA local chapters are often useful in bringing you into contact with a wider range of development people, some of whom may be looking for a writer for their projects. The trick, of course, when meeting such people is not to be too eager. Certainly you should ensure that they know you are a writer and available for work, but like any networking you should approach it carefully so that you do not alienate potential clients.

The script agency

Script agencies could be an excellent way for the game writer to find clients. However, only a few agencies currently exist, so until there are more it is difficult for the writer to get onto their books. If you are fortunate enough sign up with an agency, having someone else chase up work on your behalf and manage the contract negotiations is a real blessing. Do not sign an exclusive deal with the agency, though, as you may find yourself losing out on work that comes directly to you. Unless the agency can guarantee continuous work, expecting you to sign an exclusive deal is unfair on you.

This last point also means that you must continue to look for work yourself at the same time that the agency is doing so, but having two approaches dedicated to the task increases your chances of finding the work.

Legal documents

Something you should always be aware of when discussing projects with a development studio is the signing of a non-disclosure agreement (NDA). Although the main intention of the NDA is to protect the ideas of the developer, it should also act as protection of your ideas, too, if they are not yet in the public domain. NDAs are a normal part of creative development in general and not simply limited to games, so the likelihood is that you may have already seen and been asked to sign one in the course of your career.

NDAs, like any legal document, should be taken very seriously. When you start working on an exciting new project, there is an incredible temptation to talk about it with others – friends and colleagues, for instance – but to do so will put you in breach of the NDA. If information about the project's details were to appear in the public domain from such discussion you could be sued for a six or seven figure sum. In a business where original ideas can give a developer a real edge, those ideas must be protected if they are to survive in a harsh business environment.

Contracts for the work are always a potential minefield, but the key is to be wary of anything that puts unrealistic expectations on your time. A contract agreement must be something that both parties are happy with or it is unlikely to be a workable arrangement. Do not just sign a contract to get the work and then realise that you cannot fit the work into the timeframe stipulated. If you are unable to deliver the work as agreed you will be in breach of the contract, which could cost you both money and reputation.

Receiving credit

Credits are important to the freelance creator as it proves to potential clients that you have experience. You should ensure that the contract you sign has a clause which stipulates the credit you will receive in the credits list for the game. Sometimes, simply being credited as 'writer' can be a little vague where game development is concerned, so having a credit along the lines of 'story and dialogue' may be better. Obviously this will be adjusted to suit your own requirements.

Many games have a huge list of credits in which various aspects of development are broken down in detail. This means that 'Story', 'Dialogue' and 'Script Editor' may all be listed separately and if so you should ensure that you are listed in each category to which you contributed.

Of course, you may not be able to get exactly the credit listing you'd like, but you should at least have a credit that lists you as a writer.

Marketing yourself

Creating the correct image as a professional games writer is important. The industry as a whole is worth billions and individual projects can have development budgets in the millions, so it is essential that you give the impression that you and your talents fit into this big money industry.

Everything that has a connection to you is potential marketing for you and your work and should reflect your professionalism at all times. Your business card, stationery, promotional leaflets, web site and weblog all speak volumes about the kind of person you are. A simple business card may be all that you require – something that states who you are, what you do and gives clear contact details – but if it looks like it has been put together from stock clip art or on a machine in the local train station, then it may be working against your intentions.

Branding

Something you should consider is whether you should to market yourself as a brand. Business consultants will tell you that it gives a more professional impression if you operate under a company name, even if that company only consists of one person – you. It can show that you are completely serious in your intentions.

You can also use your own name to combine the professional feeling with the personal, so that if I operated under the name 'Steve Ince Scripts', for example, it would not only put over what I do, but also who I am.

A colleague, thinking along similar lines, felt that I should go for something more catchy and suggested 'Ince Perfect' (from inch perfect, of course), which appealed to me in principle. However, I'm always wary of using a word like 'perfect' as I think that in doing so I could be setting myself up for a fall. What the suggestion did, though, was to kick off a train of thought which led me to the point where I branded myself under the name of 'INceSIGHT', something I felt had the potential of being quite memorable. Having made this decision I then bought the domain name, incesight.co.uk, which established that I was based in the United Kingdom for anyone that considered visiting my website.

One advantage of using branding in this manner is that it allows you to separate different aspects of your professional life, should you wish to do so.

You could be a novelist as well as a games writer and you may want to separate the two to concentrate your marketing in different directions. Of course, the alternative may be preferable – if you have established yourself under your own name and want to use that to help expand your career into games, then branding yourself differently will possibly reduce the marketing strengths you have established. I imagine that if Stephen King or J.K. Rowling wanted to get into game writing they would want to use their respective established names.

Another branding possibility to consider is to find a few like-minded writers and form some kind of collective. This does not have to be anything more formal than the creation of an umbrella 'brand' that you all market yourselves under, though if you each felt it was right you could set it up as a proper scripting agency. If each member of the group has talents that complement one another, the brand has more to offer the potential client than each writer could individually.

Build a website

Because the internet is such an important part of all our lives, especially for the games industry, it is important for the game writer to maintain a strong presence there. A game developer or publisher looking for a writer will, if they do not already have the right contacts, conduct a search online as a first step. If you do not have a strong online presence you could lose out on potential clients finding you in this way. If you have any doubt about the importance of this, I can assure you that some of my best writing jobs were as a result of the clients discovering my details when conducting an online search for game writers.

Not only should you have your own website, you should also consider obtaining your own domain name, which helps to establish your internet identity and maintain your professional image. What sort of impression will you give to the world if your site and e-mail address is provided by a cheap, or free, hosting service, which may also insist on your site carrying banner advertisements? People can be very judgemental at times and when they are being cautious about spending money on the services of a writer, will often be put off by anything which might suggest that you are not as professional as you could be. If necessary, you should get some advice on domains and hosting, either through researching the subject or by consulting an expert.

Creating a look for your site that is visually appealing is important, too. Games are a very visual medium, so if a developer visits your website you do not want them to turn away because the layout suggests that you have no appreciation of the importance of visual impact. That is not to say that you need to overload your site with lots of graphics, but even a minimalist design should have a structure to it. A site where the text expands to fill the width of the screen becomes very difficult to read when the site is viewed full-screen at high resolutions – this is why newspapers and magazines have text printed in columns.

If you do not have the graphical or technical skills necessary to create the web pages yourself, you may have to pay someone to create the site for you, but it is going to be a wise business investment if the end result is something that tells the visitor that this writer has a quality site that helps establish your professional status. An alternative could be to find someone you can trade skills with. It may be that you know of or could find a graphic designer who will build your site if you do some writing work in exchange – create some dynamic copy for another client of theirs, say.

Keywords are an important part of how search engines find your site and are embedded in the information on each page. If you do not know how this is done, ensure that the person putting together your website includes all the key words you can think of that are relevant to you and your work. Words you could include are: writer, scriptwriter, video game, computer game, 'your name', experienced, etc.

Further internet presence

Having a website is all very well, but you need to draw people to it. While searching may throw up links to your site, you should also consider other ways to be more proactive in drawing visitors.

Subscribing to press release sites is an excellent way of not only keeping abreast of game news announcements, but also of making your own announcements in the form of press releases. Because many people in the industry subscribe to these sites, regular press releases are a good way of keeping your name in the eyes of potential clients, particularly if you are allowed to make announcements about big projects you have landed.

Another way of marketing yourself is to consider writing articles or columns for gaming websites. It is rare that you will be paid for such work and when you do it is more than likely to be a token payment, but you should never consider it to be a source of income but as a marketing technique. Ensure that each time you do this the page the article appears on also carries a brief bio as well as a link to your website. If you do not mind risking spam, you could also include a contact e-mail address – this ensures that the reader of the article can contact you easily. The value of writing articles can again be demonstrated by my own experience – the opportunity to write this book came about as a direct result of the commissioning editor reading one of my online articles. Links to these articles should also be placed on your website so visitors can see for themselves that you take your craft seriously.

Interviews about your writing can be a very useful way of putting over your ideas and experience in a more informal manner than writing articles and also have the benefit of making you respond to questions that you may never have thought of. Many of the interview requests I have received have been from fan sites (being a developer as well as a writer means that it is important for me to have a connection to my potential customers) but I have also done them for gaming news sites and for writers' newsletters. Sometimes they are done in conjunction with the developer you are working for, in which case the emphasis will be on that project and act as a marketing piece for the game, too. All interviews can have long term value if you approach them professionally and link to them from your website.

Creating your own weblog (or blog) is also a way of maintaining a regular presence online. By posting your thoughts and comments on writing and gaming – linking to other sites, news items, articles, reviews, etc. – you encourage other people to visit your site regularly and link to it from their own. Some people believe a blog should be updated every day, but it is better to update less regularly and ensure that what you write is of interest. Blogs are not for everyone; if you become a blogger it may take a little time to settle into a suitable style, but it should be one that reflects your interests and skills.

Non-internet marketing

Games conventions and conferences are a valuable place to mix with developers, publishers and others involved in the industry, particularly if you can get together with other game writers and share those experiences that are not currently covered by NDAs. Sometimes conferences are a valuable place to see the latest technology or demonstrations of up and coming game releases, but their real value to a freelance writer lies in the opportunities to network.

Although the internet is incredibly valuable, being able to meet people in the flesh has no substitute. Talking directly to developers gives both parties the opportunity to weigh each other up and to get a sense of whether they will be able to work together.

Many people who attend conferences have a very tight schedule, so it is not always possible to find people to talk to on the day. Try to work out a schedule of your own and arrange brief introductory meetings in advance. Call up the developer or publisher, find out who will be attending the event and make an appointment with them. Only ever ask for ten minutes of their time – if you cannot sell yourself in ten minutes you may need to work harder on pitching your skills. If the potential client is interested you can always follow up with more details after the show or you could arrange a further meeting over lunch or in the evening.

Ensure that you have a plentiful supply of business cards and a number of folders containing writing samples. However, never give away your samples unless the other person specifically asks. Not only is there the likelihood that they will get lost or left somewhere but they may fall into the wrong hands, you may also risk alienating the other person by giving them something they did not ask for. Most visitors to conventions are given a lot of stuff they carry around all day and may resent you adding to their burden.

During the week following the convention you should send out e-mails or make follow-up phone calls in which you thank the person for their time – even if they showed little or no interest – and expand on the conclusion of the meeting where relevant. Even meetings which did not go well should be followed up courteously because you rely on the network of developers and publishers, who often know each other, to make your living. Offending people or being rude to them will probably be passed around very quickly.

Keeping yourself marketable

Reading this book alone is not going to turn you into a games writer – you must also play games to understand how they work. If you do not enjoy playing games it is difficult to see how your writing skills can be applied to game development in a convincing way.

Playing a large number of full games is not always feasible for a busy writer, but as most games have a demo version, either downloadable or on a magazine cover disc, it is possible to get to play a huge variety of game styles at little or no cost. Be sure to play games on both PC and console, to understand the different approaches to the interface that are necessary to adapt to the controllers for each platform. The handheld consoles have some similarity to the larger consoles, but it is worth being aware of how they work, the limitations of the platform and how they use new features like touch sensitive screens.

The gaming industry is remarkably fast-moving and to keep yourself marketable you need to keep abreast of the developments in gaming. Visiting a number of general gaming news sites and playing as many games as possible is a must if you are to keep on top of the constant progress.

Some sites have a low signal to noise ratio and post a lot of rumour and speculation in the guise of news. Sometimes this is encouraged by the hardware manufacturers, software developers and publishers as a way of keeping their products and ongoing projects in the public eye. What it means, however, is that you may have to take much of what you read with a pinch of salt. Some of the news sites that are geared towards the business side of the industry are a little less prone to buy into the speculative hype, I feel.

Reading reviews of games can be very valuable, particularly if the site gives the visitor the chance to comment on the review or rate the game themselves. This gives you a feeling for what the game players are thinking of the games. In particular, look for reviews of games in which the story is an important part of the experience. How did the reviewer rate the story and dialogue and is there anything to be learned from what he or she said?

Being in touch with the views of game players is important in an industry where success relies on the enjoyable experience of the player relies on how they interact with the game. Subscribing to a few gaming forums can be very valuable, even if you never contribute, but much of what is posted should be treated with caution because the number of members posting is likely to be a small percentage of the total number and will possibly not always be representative of the much larger gaming audience.

4 Appendices

Design documentation

Design documentation can vary greatly from one development studio to another, therefore, what is presented here should in no way be taken as a standard format for the industry. The intention is to show the type of thing a design document may cover and how it may be laid out. Searching the internet will likely give you further design documents to read and compare.

Design document

The following is a document which outlines the design of the first section of my forthcoming game. You should note that the logic outline presented within is simply the optimal route. The player may choose to try things in a very different order and the gameplay and dialogue will reflect this as the game is implemented.

Juniper Crescent – The Sapphire Claw

Section Design Document

Section One – The Serpent's Eyes, Part One

Actors

- Scout (player character)
- Lincoln (male monkey)
- Amber (female monkey)
- Crusty (crab)
- Blinky (mouse)
- Nigel (skull)
- Skull Rock

General location notes

This section highlights aspects of locations that are important for implementation and art design. The information contained in here is necessarily brief; concept sketches of the locations and relevant maps will be included later.

1. **Intro**
    1.1. This is a mood-setting location, but also contains some gameplay.
    1.2. A **Hollow Log** can be interacted with.
    1.3. It contains a **Mouse (Blinky)** who can be talked to once he is out of the log. He will also go into the inventory.
    1.4. An **exit** on the right of the location leads to **Skull Rock 1**.

2. **Skull Rock 1**
    2.1. The location is a long one, horizontally, that scrolls left and right – **Skull Rock** can be seen in the background.
    2.2. A **Skull** has been stuck on top of a pole – it talks.
    2.3. An **exit** behind and to the right of the skull leads down into **The Valley**.
    2.4. An **exit** to the left of the location leads back to the **Intro**.
    2.5. An **exit** to the right of the location leads to **Lincoln**.

3. **Lincoln's Tree**
    3.1. A **Monkey (Lincoln)** sits up in a tree.
    3.2. A **Vine** hangs from the tree out of reach from the ground.
    3.3. An **exit** to the right of the tree leads down into **The Valley**.
    3.4. An **exit** on the left of the location leads to **Skull Rock 1.**

4. **The Valley**
    4.1. A **ravine** with a fast-flowing river cuts the location in half.
    4.2. An old, **dead tree** stands on the edge of the ravine.
    4.3. A **small rise** with a **large rock** on it is in line with the tree.
    4.4. A **Monkey (Amber)** sits on the rock.
    4.5. **Dense growth** bars the **exit** to the **Coconut Trees**.
    4.6. At the far side of the ravine, **Steps** lead up to **Skull Rock 2**.
    4.7. Also, a **path** leads down to **The Beach**.
    4.8. An **exit** on the left leads up to **Skull Rock 1**.
    4.9. An **exit** on the right leads up to **Lincoln's Tree**.

5. **The Coconut Trees**
    5.1. A small clearing with **rocks** on two sides and the **ravine** on a third.
    5.2. The fourth side is trees and an **exit** to **The Valley.**
    5.3. A stand of **Coconut trees** is set off-centre.
    5.4. A **large rock** sticks out of the ground beneath the coconut trees.

5.5. Two **moderate sized rocks** lie on the ground to one side. (??)
5.6. A **log** lies on the ground.
5.7. Lying against the rocks is an **old skeleton**.

6. **The Beach**
    6.1. A beautiful stretch of beach with a waterfall at one end and a rocky promontory at the other.
    6.2. A **path** leads off to **The Valley**.
    6.3. A large **crab** stands guard at the end of the path.
    6.4. A hole to one side of the crab serves as its hiding place.
    6.5. A number of different sized rocks lie near a rock pool.

7. **Skull Rock 2**
    7.1. The **skull** fills the height of the screen with only the top few steps visible at the bottom of the screen.
    7.2. The **steps** lead down to **The Valley**.
    7.3. The **entrance** to Skull Rock is barred by a **heavy iron bars**.
    7.4. To one side of the entrance is a **torch**.
    7.5. To the other side is the **balance puzzle**.

Logic

This section details the walkthrough logic for the section. It does not include all incidental interactions that may be added but are not required for completion of the section. Incidental interactions will be completed once the artwork has been finalised.

Intro

Opening cut scene

1. We fade up to find that Scout is emerging from the darkness of jungle-like trees. It's a moonlit night, but the light doesn't penetrate into the depths of the forest.
2. A brief voiceover explains how Scout arrived on Skull Island.

Gameplay

3. There is a hole in a hollow log that lies on the ground. Examine this and Scout comments that there's something inside, but he can't reach it.
4. Examine various background objects.
5. Exit to the Skull Rock 1 location.

Skull Rock 1

6. Once out of the trees Scout finds he's on the top of a low hill overlooking a small valley. Across the other side of the valley is Skull Rock.
7. Scout takes in his surroundings.

Gameplay

8. Talk to the Skull. Though he's been there for a long time, he's not very helpful. He's supposed to warn off intruders, but is more interested in having a chat.
   - 8.1. Will talk about the mouse.
   - 8.2. Will give a false name.
   - 8.3. Can be removed from his pole.
     - 8.3.1. The Pole can be obtained, which goes into the inventory.
9. Take the exit on the far right and go to the Lincoln location.

Lincoln's Tree

Gameplay

10. In a tree near the top of the hill is a monkey (Lincoln). Talk to the monkey, who is more smart-mouthed than is good for him.
    - 10.1. His first response is along the lines of 'Oh my god, a talking cat!' It goes downhill from there.
    - 10.2. He's then a bit unresponsive, apart from giving sarcastic responses to anything that Scout asks.
11. Beside the monkey hangs a vine.
    - 11.1 Examine the vine or try to pick it up and you find that it's out of reach.
12. Talk to the monkey about the vine and he refuses to give it to you.
13. With nothing else to do, take the exit to the Valley.

The Valley

Gameplay

14. Examine the dead tree next to the ravine.
15. Sitting on a rock on a small rise is another monkey (Amber) – she's looking very upset. Talk to her.
    - 15.1. You find out that she's had a falling out with her boyfriend, the first monkey.
    - 15.2. You find out that the boyfriend's name is Lincoln and her name is Amber. She won't elaborate any further.

16. Try using the pole on the tree.
    16.1. You get a response about needing a much greater force than that to push the tree over.
17. Try using the pole on the rock.
    17.1. Amber will yell at you to stop.
18. Leave the Valley and return to the Lincoln's Tree location.

Lincoln's Tree

Gameplay

19. Talk to the first monkey, Lincoln, and ask him about his girlfriend and why she's angry.
    19.1. Lincoln says that Amber's furious because he won't collect the coconuts for tonight's party. He doesn't see why it always has to be him and not her.
    19.2. Lincoln still won't give the vine to Scout.
20. Return to the Intro location.

Intro

Gameplay

21. Use the pole on the hollow log.
    21.1. This causes a mouse to pop out of the hole in the top.
22. Talk to the mouse, who you will find is called Blinky.
23. Pick up the mouse and he goes into the inventory.
24. Return to the Valley.

The Valley

Gameplay

25. Go to the area behind where Amber is sitting where you will find that there is dense vegetation barring the way into the trees.
26. Use the pole on the vegetation.
    26.1 This breaks some of the branches and opens up the exit.
27. Go through the exit to the Coconut Trees.

Coconut Trees

Gameplay

28. Scout finds himself in a small area that has a few coconut trees at one side. The coconuts are way too high to reach.

29. There is a split log lying on the ground.
    29.1 Pick up the split log.
    29.2 Place it on the half-buried rock beneath the trees to form a seesaw.
30. Pick up one of the heavy rocks lying around.
    30.1. The rock can be placed on the end of the seesaw.
    30.2. Another rock can be picked up and dropped onto the other end of the seesaw.
        30.2.1. This causes the first rock to fly into the air. Although it comes close, it misses the coconuts.
        30.2.2. This can be repeated any number of times with the same result.
31. Pick up a rock.
    31.1. Put Blinky onto the see-saw.
    31.2. Use the rock on the seesaw.

Cut scene

32. Scout drops the rock on the other end of the seesaw.
33. Blinky is flung into the air with great force.
34. He smashes into the tops of the trees, hitting the coconuts with great force and the whole screen shakes.
35. Blinky asks Scout what he thinks he's doing.
36. Scout tells him to get him a coconut.
37. Blinky refuses.

Gameplay

38. Scout talks to Blinky.
    38.1. Promises not to eat him.
    38.2. Promises to take him off the island with him.
    38.3. Blinky drops down two coconuts.
39. Pick up the coconuts – they go into the inventory.
    39.1. This triggers an animation of more coconuts falling.
    39.2. Coconuts rain down on top of Scout, hitting him on the head, followed by Blinky.
40. Pick up Blinky again – he returns to the inventory.
41. There is an old skeleton in this location.
    41.1. Search the remains of the skeleton and you will find an old tinderbox, which goes into the inventory.
42. Leave this area and return to The Valley.

The Valley

Gameplay

43. Try giving the coconuts to Amber.
    43.1. She will say that it's very kind, but Lincoln's the one who should be getting them.
44. Return to Lincoln's Tree.

Lincoln's Tree

Gameplay

45. Give the coconuts to Lincoln.
    45.1. If you've already talked about the vine he will give it to you in exchange.
    45.2. Lincoln then jumps down out of the tree and rushes off.
    45.3. If you didn't previously examine or try to get the vine, Lincoln leaves it hanging on the tree.
46. Use the pole to reach the vine and free it from where it hangs.
    46.1. The Vine goes into the inventory.
47. Return to the Valley.

The Valley

Gameplay

48. The monkeys have now disappeared.
49. Now use the pole to lever the rock on the rise.
    49.1. It will roll down the incline and into the tree.
    49.2. The tree is knocked over so it makes a bridge across the ravine.
50. Cross this bridge to the other side.
51. Climb up the steps to Skull Rock 2.

Skull Rock 2

Gameplay

52. Examine the entrance (the mouth of the skull).
    52.1. Heavy iron bars block the entrance to Skull Rock.
    52.2. There is no visible means of opening up the entrance directly.
53. To one side of the door is what appears to be a set of scales, though each pan has its own pointer to mark off the weight.
    53.1. A number of different weights lie on the ground and these can be picked up and used on the pans.

53.2. A quick experiment will show that the pointer on one side moves twice as far as the pointer on the other when the same weight is placed upon it.
53.3. There is a red line that goes across the two pointer scales at the same height and the two pointers must be made to exactly reach this line in order that the door will open.
53.4. No matter how you try, you will find that you can't complete this puzzle at the moment because two of the weights are missing.

54. Leave the puzzle alone and pick up the unlit torch on the far side of the doorway.
55. Go down to the valley and then take the exit to the Beach.

The Beach

Gameplay

56. The path is blocked by a huge ferocious-looking crab.
57. Try to pass the crab and he will lunge at you, forcing you back.
58. Talk to the crab and you'll find out he's called Crusty.
59. Examine the hole that lies just to one side of Crusty.
60. Talk to Crusty about the hole.
    60.1. He'll tell you that it's his home.
    60.2. Talking to him further reveals that he's nocturnal and he hides in his home during the day to avoid the bright sunlight.
61. Plant the torch into the ground near Crusty.
62. Use the tinderbox on the torch.
    62.1. It bursts into flame and causes Crusty to scurry to his hole.
    62.2. You can now get to the beach proper.
63. Examine a nearby rock pool.
    63.1. Collect four different-sized rocks from the rock pool – these go into the inventory.
64. Return to Skull Rock 2.

Skull Rock 2

Gameplay

65. Interact with the scales.
    65.1. With a little experimentation you find which of the four rocks are substitutes for the missing weights.
    65.2. By combining the weights in the right way, both pointers can be made to exactly meet the red line and the entrance opens.
66. As the iron bars drop into the ground, the Skull Rock talks to Scout in a deep voice.

Cut Scene

67. Scout steps back, momentarily surprised.
68. The voice warns of a deadly peril if he should venture further.
69. Scout asks what the peril is, but is given no other answer.
70. Treading carefully, Scout enters the huge gaping mouth...

END OF PART ONE.

Character profile template

The following is a character profile template that I have used on a few projects to help clarify the characters, not only for my own benefit, but for others on the team such as designers, animators and character modellers. The character profile documents should be seen as living documents that are updated as required.

The format, type of information and the amount you include will depend on the game itself and on how important the character is in the game. A simple background character may not need a profile at all, for instance.

Name:

Fill in the character's name.

Gender:

What gender are they? Non-human characters may not conform to regular male/female gender roles. Some humans may not, come to that.

Race:

This may include fantasy races or alien species.

Type/Role:

What type of character is this? What is the role of the character in the gameplay and in the story? How does this character relate to other characters? What are this character's key events?

Background:

What is the character's background and family history? Was there a happy childhood, a history of abuse, extensive tragedy or hardship? What has made the character the person they are? What was their education? What has led them to the point at which the game starts?

Visual (concept art/reference):

Place here an image from the character concept art when the art department has created it.

Extraordinary traits and abilities:

What makes the character special?

Major deficiencies:

What are the character's flaws?

Favourite thing:

This could be anything from a small trinket given by a late mother, to an expensive item for which they saved towards over many years. Thinking about the character's favourite thing will really help define them in your mind.

What they like to do most:

Hobby, pastime, an aspect of their career?

Physical attributes (height, weight, build):

Describe them.

Unusual mannerisms:

Be wary of creating something just for the sake of it. Watch the mannerism doesn't come across as gimmicky.

Mode of speech – accent?

May depend on their background, race, education, etc.

What makes them laugh?

Do they laugh at kittens playing or at the misfortune of others, for instance?

What makes them cry?

Personal bereavements? The future of the planet? Poverty?

Mode of dress:

Are they particular about their appearance? Are they sloppy? Do they wear a uniform as part of their job?

Their work:

How does the character make a living?

Mood swings:

What is it that makes them quick to anger or to descend into deep melancholy? What makes them instantly happy?

Gameplay abilities:

This is likely to be something filled in by the design department and should include everything the character does in the game – shoot, run, throw grenades, etc.

Sample script

The following is a sample script taken from my forthcoming game, *Juniper Crescent – The Sapphire Claw*. This includes all the logic which controls the dialogue choices and responses as well as some function calls – to animate characters, for instance. The game is being developed using the Wintermute Engine and tool set, so the script format is the one that this tool set uses.

Talk to the Skull

```
global SkullVar;
global BlinkyVar;
global Coconut;
global InvisibleVar;

var Blinky = Scene.GetNode("blinky");
var Skull = Scene.GetNode("skull");
var Pole = Scene.GetNode("pole");
var Scout = actor;

////////////////////////////////////////////////////////////////////////////////

on "Talk"
{
GoToObject();
Game.Interactive = false;

// greetings
if(!SkullVar.Talked)
  {
    if(!SkullVar.Moved) Scout.ForceTalkAnim("actors\scout\rr\talkup.sprite");
    Scout.Talk("Hello, skull.");
    Skull.ForceTalkAnim("sprites\Skull\talk.sprite");
    Skull.Talk("Hello, cat. It's so good to meet you.");
    // set the flag, so that we know we've already talked to him
    SkullVar.Talked = true;
  }
```

```
else
{
    if(!SkullVar.Moved) Scout.ForceTalkAnim("actors\scout\rr\talkup.sprite");
    Scout.Talk("Hi, it's me again.");
    Skull.ForceTalkAnim("sprites\Skull\talk.sprite");
    Skull.Talk("Hello. I'm glad you came back to talk some more.");
}

if(InvisibleVar == true)
{
    if(!SkullVar.Moved) Scout.ForceTalkAnim("actors\scout\rr\talkup.sprite");
    Scout.Talk("You can see me?");
    Skull.Talk("Yes, of course I can.");
    if(!SkullVar.Moved) Scout.ForceTalkAnim("actors\scout\rr\talkup.sprite");
    Scout.Talk("But I'm invisible.");
    Skull.Talk("Not to me, you're not.");
  }

  // and let the dialogue choices begin
  SkullTalk1();

  // restore interactivity
  Game.Interactive = true;
}

////////////////////////////////////////////////////////////////////////////////

on "LookAt"
{
  Scout.GoToObject(this);
  Game.Interactive = false;

  if(!SkullVar.Moved) Scout.ForceTalkAnim("actors\scout\rr\talkup.sprite");
  Scout.Talk("This is a sorry-looking skull.");
  Skull.ForceTalkAnim("sprites\Skull\talk.sprite");
  Skull.Talk("Don't be so rude. You can hurt a person's feelings saying things
  like that.");
  if(!SkullVar.Talked)
  {
```

```
    Scout.ForceTalkAnim("actors\scout\rr\talkup.sprite");
    Scout.Talk("Good grief, a talking skull!");
    // set the flag, so that we know we've already talked to him
    SkullVar.Talked = true;
  }
  Game.Interactive = true;
}

////////////////////////////////////////////////////////////////////////////////

on "Take"
{
  Scout.GoToObject(this);
  Game.Interactive = false;

  if(!SkullVar.Moved)
  {
if(!SkullVar.Moved) Scout.ForceTalkAnim("actors\scout\rr\talkup.sprite");
    Scout.Talk("Excuse me a moment.");
    Skull.ForceTalkAnim("sprites\Skull\talk.sprite");
    Skull.Talk("Hey, just what do you think you're doing?");
    Skull.SkipTo(440, 450);
    Pole.Interactive = true;
    SkullVar.Moved = true;
  }
  else
  {
    Scout.Talk("No need to move it again.");
  }
  Game.Interactive = true;
}

////////////////////////////////////////////////////////////////////////////////

on "Cards"
{
  Scout.GoToObject(this);
  Game.Interactive = false;
```

```
  if(!SkullVar.Moved) Scout.ForceTalkAnim("actors\scout\rr\talkup.sprite");
  Scout.Talk("Would you like to play a game of cards.");
  Skull.ForceTalkAnim("sprites\Skull\talk.sprite");
  Skull.Talk("And just how do you expect me to hold my cards?");

  Game.Interactive = true;
}

////////////////////////////////////////////////////////////////////////////////

on "Coconut"
{
  Game.SelectedItem = "null";
  Scout.GoToObject(this);
  Game.Interactive = false;

  Scout.Talk("Hey, Coconut. Come over here, will you?");
  Coconut.Talk("Okay...");
  Coconut.GoTo(551, 565);
  Coconut.Talk("What do you want?");
  Scout.Talk("Have you ever seen a talking skull before?");
  Coconut.Talk("No...");
  if(!SkullVar.Moved) Scout.ForceTalkAnim("actors\scout\rr\talkup.sprite");
  Scout.Talk("Hey, skull, say something cool.");
  Skull.ForceTalkAnim("sprites\Skull\talk.sprite");
  Skull.Talk("I'm not a trained parrot, you know.");
  Coconut.Talk("Hey, that's neat.");

  Game.Interactive = true;
}

////////////////////////////////////////////////////////////////////////////////

on "LeftClick"
{
  Scout.GoToObject(this);
}
```

```
//FUNCTIONS

/////////////////////////////////////////////////////////////////////////////////

function GoToObject()
{
  Scout.GoTo(300, 600);
  Scout.TurnTo(DI_DOWNRIGHT);
}

/////////////////////////////////////////////////////////////////////////////////

function SkullTalk1()
{
  var Responses;
  var Selected;
  var Loop = true;

  while(Loop)
  {
     // prepare the choices
     Responses[0] = "Talking skull";
     Responses[1] = "Name";
     Responses[2] = "Pole";
     Responses[3] = "Skull Rock";
     Responses[10] = "Exit";

     // fill the response box
     if(!SkullVar.Talking) Game.AddResponse(0, Responses[0]);
     if((!SkullVar.Name)||((BlinkyVar.Nigel)&&(!SkullVar.Nigel)))
     Game.AddResponse(1, Responses[1]);
     if(!SkullVar.Pole) Game.AddResponse(2, Responses[2]);
     if(!SkullVar.SkullRock) Game.AddResponse(3, Responses[3]);
     Game.AddResponse(10, Responses[10]);

     // let the player choose one
     Selected = Game.GetResponse();
```

```
// now let the conversation develop depending on the selected sentence
if(Selected==0) //Talking Skull
{
   if(!SkullVar.Moved)
   Scout.ForceTalkAnim("actors\scout\rr\talkup.sprite");
   Scout.Talk("I never expected you to talk.");
   Skull.ForceTalkAnim("sprites\Skull\talk.sprite");
   Skull.Talk("Why not? You seem to be doing remarkably well for a pussy-
   cat.");
   if(!SkullVar.Moved) Scout.ForceTalkAnim("actors\scout\rr\talkup.sprite");
   Scout.Talk("But you're just a skull.");
   Skull.ForceTalkAnim("sprites\Skull\talk.sprite");
   Skull.Talk("So, you've got something against skulls have you?");
   if(!SkullVar.Moved) Scout.ForceTalkAnim("actors\scout\rr\talkup.sprite");
   Scout.Talk("No, I...");
   Skull.ForceTalkAnim("sprites\Skull\talk.sprite");
   Skull.Talk("It's always the same - prejudice just follows me around.");
   if(!SkullVar.Moved) Scout.ForceTalkAnim("actors\scout\rr\talkup.sprite");
   Scout.Talk("Really?");
   Skull.ForceTalkAnim("sprites\Skull\talk.sprite");
   Skull.Talk("Well, not that I actually go anywhere...");
   Skull.ForceTalkAnim("sprites\Skull\talk.sprite");
   Skull.Talk("I blame Ray Harryhausen for stereotyping the whole subject
   of skeletons.");
   SkullVar.Talking = true;
}

else if(Selected==1) //Name
{
   if (!SkullVar.Name)
   {
      if(!SkullVar.Moved) Scout.ForceTalkAnim("actors\scout\rr\talkup.
      sprite");
      Scout.Talk("What's your name?");
      Skull.ForceTalkAnim("sprites\Skull\talk.sprite");
      Skull.Talk("I'm not in the habit of giving my name out to just anyone,
      you know.");
      if(!SkullVar.Moved)  Scout.ForceTalkAnim("actors\scout\rr\talkup.
      sprite");
```

```
  Scout.Talk("It's something embarrassing, isn't it?");
  Skull.ForceTalkAnim("sprites\Skull\talk.sprite");
  Skull.Talk("Not at all! It's, er... Cutthroat Jake! Yes.");
  if(!SkullVar.Moved)  Scout.ForceTalkAnim("actors\scout\rr\talkup.
  sprite");
  Scout.Talk("You just made that up.");
  Skull.ForceTalkAnim("sprites\Skull\talk.sprite");
  Skull.Talk("No I didn't!");
  Skull.ForceTalkAnim("sprites\Skull\talk.sprite");
  Skull.Talk("You can call me Jake if you like.");
  SkullVar.Name = true;
  Skull.Caption = "Jake";
}

if ((BlinkyVar.Nigel)&&(!SkullVar.Nigel))
{
  if(!SkullVar.Moved)  Scout.ForceTalkAnim("actors\scout\rr\talkup.
  sprite");
  Scout.Talk("That mouse over there...");
  Skull.ForceTalkAnim("sprites\Skull\talk.sprite");
  Skull.Talk("I hate that mouse. Do you have any idea what he did to
  me?");
  if(!SkullVar.Moved)  Scout.ForceTalkAnim("actors\scout\rr\talkup.
  sprite");
  Scout.Talk("I...");
  Skull.ForceTalkAnim("sprites\Skull\talk.sprite");
  Skull.Talk("He had the audacity to build a nest in my eye socket.");
  Skull.ForceTalkAnim("sprites\Skull\talk.sprite");
  Skull.Talk("I was only able to drive him out by whistling for six hours
  straight.");
  Skull.ForceTalkAnim("sprites\Skull\talk.sprite");
  Skull.Talk("Do you realise how difficult that is when you don't have
  any lips...?");
  if(!SkullVar.Moved)  Scout.ForceTalkAnim("actors\scout\rr\talkup.
  sprite");
  Scout.Talk("The mouse told me your real name is Nigel.");
  Skull.ForceTalkAnim("sprites\Skull\talk.sprite");
  Skull.Talk("Well...");
```

```
        if(!SkullVar.Moved)  Scout.ForceTalkAnim("actors\scout\rr\talkup.
        sprite");
        Scout.Talk("What sort of a name is Nigel for a skull?");
        Skull.ForceTalkAnim("sprites\Skull\talk.sprite");
        Skull.Talk("I know, I know...");
        SkullVar.Nigel = true;
        Skull.Caption = "Nigel";
        }
}

else if(Selected==2) //Pole
{
    if(SkullVar.Moved)
    {
        Scout.Talk("Why were you stuck on that pole?");
        Skull.ForceTalkAnim("sprites\Skull\talk.sprite");
        Skull.Talk("I was placed there as a warning to intruders.");
        }
    else
    {
        if(!SkullVar.Moved)  Scout.ForceTalkAnim("actors\scout\rr\talkup.
        sprite");
        Scout.Talk("Why are you stuck on that pole?");
        Skull.ForceTalkAnim("sprites\Skull\talk.sprite");
        Skull.Talk("I've been placed here as a warning to intruders.");
    }
    if(!SkullVar.Moved) Scout.ForceTalkAnim("actors\scout\rr\talkup. sprite");
    Scout.Talk("Intruders?");
    Skull.ForceTalkAnim("sprites\Skull\talk.sprite");
    Skull.Talk("People like yourself, who come here looking for the treas-
    ure hidden in Skull Rock.");
    Skull.ForceTalkAnim("sprites\Skull\talk.sprite");
    Skull.Talk("Oops! Forget I ever said that.");
    if(!SkullVar.Moved)     Scout.ForceTalkAnim("actors\scout\rr\talkup.
    sprite");
    Scout.Talk("Not very good at your job are you?");
    Skull.ForceTalkAnim("sprites\Skull\talk.sprite");
    Skull.Talk("You call this a job? It's certainly not one I would have chosen
    myself, if I'd had any say in the matter.");
```

```
    Skull.ForceTalkAnim("sprites\Skull\talk.sprite");
    Skull.Talk("It's not as if I can just walk away and get another job.");
    Skull.ForceTalkAnim("sprites\Skull\talk.sprite");
    Skull.Talk("When a skull doesn't even have the rest of his skeleton with
    him, career options are severely limited.");
    if(!SkullVar.Moved) Scout.ForceTalkAnim("actors\scout\rr\talkup.sprite");
    Scout.Talk("Talk a lot, don't you?");
    Skull.ForceTalkAnim("sprites\Skull\talk.sprite");
    Skull.Talk("It's the sheer and utter boredom – I haven't had a decent
    conversation in forty three years.");
    Skull.ForceTalkAnim("sprites\Skull\talk.sprite");
    Skull.Talk("Listen, you won't tell anyone that I mentioned the treasure,
    will you?");
    SkullVar.Pole = true;
}

else if(Selected==3) //Skull Rock
{
    if(!SkullVar.Moved) Scout.ForceTalkAnim("actors\scout\rr\talkup.sprite");
    Scout.Talk("What can you tell me about Skull Rock?");
    Skull.ForceTalkAnim("sprites\Skull\talk.sprite");
    Skull.Talk("Not very much, I'm afraid. I've never even seen it.");
    if(!SkullVar.Moved) Scout.ForceTalkAnim("actors\scout\rr\talkup.sprite");
    Scout.Talk("Why, because you don't have any eyes?");
    Skull.ForceTalkAnim("sprites\Skull\talk.sprite");
    Skull.Talk("No, I can see you just fine.");
    if(!SkullVar.Moved) Scout.ForceTalkAnim("actors\scout\rr\talkup.sprite");
    Scout.Talk("Weird, that...");
    Skull.ForceTalkAnim("sprites\Skull\talk.sprite");
    Skull.Talk("Ever since I was placed here I've been facing in the same
    direction.");
    Skull.ForceTalkAnim("sprites\Skull\talk.sprite");
    Skull.Talk("And the local monkeys are really awful – telling me how
    beautiful the view is and how wonderful the sun looks as it sets over the
    sea.");
    if(!SkullVar.Moved) Scout.ForceTalkAnim("actors\scout\rr\talkup.sprite");
    Scout.Talk("Now, don't get upset.");
    Skull.ForceTalkAnim("sprites\Skull\talk.sprite");
```

```
        Skull.Talk("I'd weep if I had any tear ducts...");
        SkullVar.SkullRock = true;
        // go to the second branch of dialogue
        //SkullTalk2();
      }

      else if(Selected==10) //Exit
      {
        if(!SkullVar.Moved) Scout.ForceTalkAnim("actors\scout\rr\talkup.sprite");
        Scout.Talk("I'll be seeing you.");
        Skull.ForceTalkAnim("sprites\Skull\talk.sprite");
        Skull.Talk("Okay, bye. Don't forget to come back soon.");
        Loop = false; // we want to end the dialogue
      }
    }
  }
```

Useful reading, web sites and games to play

The following is a list of resources that I found to be very useful, stimulating, insightful, a great resource or great fun. A complete list would be huge, so consider this to be edited highlights.

BOOKS

Robert McKee, *Story*. Methuen, 1999.

Rib Davis, *Writing Dialogue for Scripts*. A&C Black (Publishers) Limited, 1999.

William Strunk jr. and E.B. White, *The Elements of Style* Fourth Edition. Longman Publishing, 1979.

J. Michael Straczynski, *The Complete Book of Screenwriting*. Titan Books, 1997.

François Dominic Laramée, Editor, *Secrets of the Game Business* Second Edition. Charles River Media, 2005.

MAGAZINES

Develop, Intent Media. Published monthly. www.developmag.com

MCV, Intent Media. Published weekly. www.mcvuk.com

WEB SITES

International Game Developers Association, www.igda.org

Game Writers' Special Interest Group, www.igda.org/writers

Eurogamer, www.eurogamer.net

Digiplay Initiative, www.digiplay.org.uk

WomenGamers.Com, www.womengamers.com

Gamasutra, www.gamasutra.com

Games Press, www.gamespress.com

GamesIndustry.Biz, www.gamesindustry.biz

Wintermute Engine, www.dead-code.org/index2.php/en

Game Maker, www.gamemaker.nl

All Game Guide, www.allgame.com

The Escapist, www.escapistmagazine.com

GAMES TO PLAY

Day of the Tentacle, LucasArts Entertainment Company, 1993.

Broken Sword – The Shadow of the Templars, Revolution Software, 1996.
Grim Fandango, LucasArts Entertainment Company, 1998.
The Longest Journey, Funcom, 2000.
Metal Gear Solid, Konami, 2000.
Vagrant Story, Squaresoft, 2000.
Final Fantasy IX, Squaresoft, 2001.
Half-Life 2, Valve, 2004.
The Moment of Silence, House of Tales, 2004.
Beyond Good and Evil, Ubisoft, 2004.
Lego Star Wars, LucasArts, 2005.
Another Code: Two Memories, Nintendo, 2005.
Psychonauts, Double Fine Productions, 2006.

Glossary

The following is a list of terms used in game development with which you may be unfamiliar. This is not intended to be an exhaustive list, but hopefully it will help you to understand gaming terminology.

Standards within the industry are highly variable, so it could be that the company you work with has their own take on these definitions and may even have different terms altogether.

Alpha: The stage in development where everything in the game is in place, but may still contain a large number of bugs and may need polishing in certain areas.

Anim: An animation file. Because games must react to the player, most things are parcelled up into discrete units. Anims are small units of animation that can be triggered very quickly as required.

Artificial intelligence (AI): In games this generally has a different meaning from true AI, where the artificial intelligence is expected to learn from the input of the user. In games, AI is usually a pre-defined set of behaviours or scripts that are applied to a character at run time to give the impression that they are reacting intelligently to the actions of the player.

Audio design: How music and sound effects will work in the game, these may include sounds based on the position of objects in the game world or multi-layered music which is context sensitive.

Avatar: Another term for the player character – the character the player controls in the game world. The term is more commonly used in online games than single player games.

Background characters: Characters that may serve no purpose other than scenery to make a world feel more alive. Some background characters are interactive but usually speak very generic lines. In action games they may also be used as cannon fodder.

Beta: The stage in development where all the final resources are in the game and the final phase of testing begins in the lead up to release.

Boolean variable: A variable that can either be true or false. Very useful for keeping track of imparted information in conversation scripts.

Branching: Where the gameplay and/or story can take different routes depending on the choices of the player.

Breadcrumbing: To ensure that the player does not become lost in a potentially bewildering game world, conversational clues can act as breadcrumbs to guide the player towards their objectives.

Collectable items: Items in the game world that the player can collect and which usually go into an inventory. They may be swords, armour, laser pistols, magical beads, rubber ducks, etc. Often they are acquired by achieving goals and can be traded for other items or for the game's currency. They may be used to solve puzzles or overcome other obstacles.

Comment: Often a voiceover that the main character will speak as if to themselves or to the player about the game world or the objects in it with which they interact. This is also a term used in writing code or logic scripts and refers to any explanatory text which does not get compiled into the game.

Conditional dialogue: When an interactive scene unfolds, some lines of dialogue may only play out if the right conditions are met. For example, this may depend on the character having witnessed an event or obtaining an important item.

Console: A video game system that generally connects to your television set to display the games. Also known as game console. They are generally regarded as different entities to personal computers.

Critical objectives: The objectives the player must complete if they are to reach the end of the game. Non-critical objectives are optional for the player and are often known as side quests. The two types of objectives are not necessarily distinguished in the game.

Critical path: A route through the game which the player must follow to complete the game. If a game includes branching, there may be multiple critical paths. This is also a term used in game development scheduling in which all the tasks on the critical path have the largest impact on the schedule if they are delayed in any way.

Cut scene: This is a term used for when the game cuts away from the interactive and displays a pre-defined scene or sequence of scenes. The purpose of the cut scene is usually to advance the story through the revealing of plot elements. Cut scenes were often pre-rendered and displayed as FMV, though they are increasingly using the game's run-time engine.

Deliverables: The tasks to be completed by a specific date in order to meet the conditions of a scheduling and/or contractual milestone. These may include story completion, level completion, character animations complete, etc.

Design document: This is the complete description of all the details relevant to the creation of the game. It may not be a single document, but a collection that covers gameplay mechanics, visual style, technical information, story and character information and level designs. This can also be known as the game design.

Design overview: A document which summarises all the gameplay elements and level designs without going into all of the technical detail.

Developer: The team responsible for making the game. The size can vary from single individuals (generally independent developers) upwards, with some teams consisting of a hundred or more artists, animators, programmers, game designers, writers, etc. The development studio may have a large number of staff, but may split into smaller teams working on a number of parallel projects.

Dialogue engine: How the characters speak to one another and how the player interacts during conversations is controlled by the dialogue engine. The design team may consult with the writer to help define how it works.

Dialogue script: Strictly speaking, this is the document containing the dialogue as written by the game writer. The format may vary depending on how the developer and the writer define it, based on the requirements of the development tools.

Edutainment: These are games that are educational in nature. Teaching while having fun.

Environment: A section of the game world. Because of memory constraints of gaming systems, the game world is split up into a number of environments. These are generally modelled in 3D these days, though some 2D environments are still created using hand-drawn backgrounds or 3D environments that have been pre-rendered into 2D.

Establishing shot: An introductory scene. This may be used for mood purposes and may be used for the whole game and/or to introduce each of the levels. May also be used to put the player character into the context of his environment and help the player see the objectives.

Event: Whenever a game object moves from one state to another, this is said to be an event. The triggering of events can also happen when objects (characters, say) cross pre-defined lines, or move into certain areas of the game world. Boolean variables are usually used to keep track of whether important events have been triggered.

Export tool: The game's data may well be stored in a way that is not ideal for other purposes, so an export tool takes the relevant data and presents it in a more useful format. Game scripts can be very complex and would be unreadable for actors trying to concentrate on their lines, so an export tool would take out all of the detail the actors do not need to see and presents the scripts in a readable manner.

Flag: Another term for boolean variable. Used in the sense of 'raising a flag' to indicate that an event has happened.

FMV: This is short for full motion video and originates from a time when games moved from floppy discs to CD format. FMV is a pre-rendered sequence that spools off the disc or the hard drive (if installed there) and essentially plays like a small film. FMV is generally being phased out, for the sake of visual consistency, as game engines become more powerful.

Game design: The act of designing the gameplay of the game and anything connected with it. Game design should be seen as separate from visual or graphic design, but they may affect one another. The game design can also be another term for the design document.

Game designer: The person who creates the game design for the game project. For a large project, there may be a number of designers, possibly consisting of lead and/or senior designers, junior designers, level designers. If a game designer also writes dialogue and the game's story, they may be classed as a writer-designer.

Gameplay: The interactive nature of the game and the obstacles the players must overcome as they work towards the game's objectives.

Gameplay mechanics: How the gameplay works – the rules, the interface, the scoring, etc.

Gameplay nodes: Places where the player must pass through for the game to progress. The player may be forced to make choices that open up new nodes or close down others, or it may be something simple like finding the way to open the secret door. Whatever else might have happened before passing through the node – shooting zombies or collecing gold rings – it is the passing the obstacle of the gameplay node that enables the player to move forward.

Game writer: A person who writes or contributes to the story, dialogue, characters or back story for a game. A game writer may also act as a lead writer where there is a team of writers on a project. May also act as script editor to provide consistency over an extensive game.

Gold master: The final version of the game that is ready for release. In theory, the gold master should be free of any bugs or other flaws.

Heads up display: Any information that is displayed on screen that is aimed specifically at the player and not really part of the game world. This can include such things as positional maps, health meters, inventory and current objective. Exactly what is displayed and the style it is displayed in will vary greatly from game to game.

Immersive gameplay: When the game offers a pleasurable experience in which the players lose themselves in the game world to the exclusion of everything around them.

Interactive narrative: The story, or story elements, is affected by the actions or choices of the player. This may lead to multiple endings or give multiple paths to flavour the same ending.

Interactive plot: This is slightly different to the interactive narrative in that the story is fundamentally the same, though the player has control over how or in what order that story is revealed.

Joypad: The control device that connects to a console. Versions to connect to computers are also available.

Level: A section of gameplay that's separated from other such sections in some way. Originally came from games in which an increase in level meant that the gameplay got a little harder. It can also refer to the level a character has achieved in a game, particularly RPGs where characters increase their level as they earn experience points.

Level design: This is a section of the game design which concentrates on a single section, or level, of the game. This may be created from a broad design document by the game designer or by a specific level designer acting under a lead game designer.

Level designer: A person whose job it is to create detailed level designs from an overview design.

Linear: When a game has no branching and all of the game levels are played in a fixed order, it is said to be linear.

Localisation: This is the act of creating versions of the game in languages other than that in which it was originated. Localisation may involve some language specific graphics, translation of the game's text, specific voice recording and specific version testing.

Location: Another term for environment.

Logic script: This is a script that is usually written in a high level code and deals with all of the logic in the game. It is usually broken down into a series of smaller scripts. Many logic scripts in a story driven game will contain the dialogue itself and the logic that drives it, as well as keeping track of the progress of the player through the game and story.

Logline: A single sentence that is designed to summarise the game in a dynamic way. A hook to get the potential customer interested in discovering more.

Milestone: A scheduling and contractual term, it defines a point in time at which pre-agreed deliverables will be completed. Milestones are a way of measuring the progress of the game's development and if a publisher is funding this, each milestone will probably have a percentage of that funding payable on completion of the milestone.

Model: Usually refers to a 3D character model that appears in the game, though it can also refer to a 3D location or environment model.

Narrative design: This is a broad document which looks at the high level design of the story. Though it will contain the flow of the story, cover the main characters and how they interact, it will also tie the story into the flow of the game and the broad Game Design.

Next-gen: A term used to denote the next generation of hardware and gaming software. Hype surrounding the proposed launch of a new console can start building up as much as two years before it appears.

Non-linear: Any game in which the player has a degree of choice in which order they play through the various elements or levels is regarded to be non-linear.

Non-player character (NPC): Any character in a game that is not controlled by the player. All their actions and responses are controlled by the scripts or AI from within the game engine.

Objective screen: The different interface images are referred to as screens, so the quest screen or objective screen is one which displays the player's current objectives or the quests (missions in some games) that the player character has undertaken. These screens are usually hidden until activated by a specific key or button press or chosen from an in-game menu.

Outsourcing: Similar to sub-contracting, outsourcing takes place when aspects of the game's development are handled by individuals or companies outside of the development studio. Translation and recording are typically outsourced, but graphics, animation and dialogue are increasingly outsourced, too.

Parallel streaming: When a game has story or gameplay paths that run parallel to each other and the player is able to switch between them when they like.

Player character (PC): The character the player controls when playing the game. The term avatar is becoming an increasingly popular alternative, particularly as the abbreviation, PC, can be confused with personal computer.

Plot: How the story is revealed to the player through the events and information that is shown through the interaction with the game world and the objects and characters within it.

Pre-rendered: A sequence or environment which is created in its final form prior to the game being compiled. It is then displayed in the game exactly as created.

Proposal: A document, or set of documents, that is used to sell the game concept to the publisher. Proposals can consist of hundreds of pages of story, design, concept art and technical information, but will also include a short Synopsis section.

Publisher: A company that manufactures, distributes and markets the game created by the developer. The publisher may also provide funding for the development of the game. Some publishers also have their own internal development studios.

Quality assurance (QA): The testing of the game to ensure that it works as intended. Approval of the game for it to be released on the consoles is dependent on it successfully meeting the rigorous demands of that particular console's QA department.

Real time: Graphics that are put together at the time of playing the game from 3D information that has been loaded into memory is regarded as real time rendering. Also, gameplay which continues, regardless of the player's input, is said to be real time gameplay.

Recording script: The dialogue for the game formatted into a series of documents from which the actors can easily read in the recording studio.

Replay value: Also known as replayability. The amount of encouragement a game gives to the player to replay the game. This can be in the form of unlockable content, different outcomes, variable gameplay, etc.

Script: This can mean a number of things – the document the writer creates which contains the game's dialogue; the logic for the game; the high-level coding language used to create the logic; the actual creation of the script (as in, to script the dialogue).

Section design: Another term for level design.

Sequence: A series of scenes and/or visuals that play out at a key point in the game, often triggered by the actions of the player. A cut scene may also be known as a sequence. A sequence may also refer to a series of gameplay events.

Side quest: Although the completion of a side quest is not necessary for the player to complete the game, doing so can add greatly to the overall experience. Side quests help make the world seem a larger, more vibrant place. Side quests may give the player valuable rewards, bonuses and even additional weapons or playable characters.

Sprite: This is a 2D graphical element displayed on a game screen. In a 2D game, sprites are used to represent everything from the characters moving about to background objects to the items the player collects. Even in a 3D game, sprites may be used to represent inventory items or part of the heads up display.

Storyboard: A series of sketches or other visuals that help represent the flow of the game, timing of animations, the cinematic visuals of a cut scene.

Subplot: A secondary plot which complements or conflicts with the main plot to add richness or additional drama. Some subplots may be tied in with side quests and be entirely optional.

Synopsis: This may be an outline of the game's story, but may also be an outline of the whole of the game, particularly when it is part of a proposal.

Target audience: The section of the game playing population for which the game will have the maximum appeal.

Testing: The thorough playing of the game over and over again to identify any gameplay issues, technical problems, graphical glitches or writing inconsistencies. Testing is often more formally referred to as quality assurance these days.

Transition: The act of changing from one state to another, which triggers an event. This may be tied into the story/plot, gameplay actions or real time engine activity.

Translation script: A version of all of the dialogue and other in-game text that requires translation into other languages. This may be formatted very differently to the recording script and is often presented in a spreadsheet or database file.

Treatment: A high-level document which outlines the intentions of the game and what it will offer to the player in terms of originality and excitement. There may be some overlap with a game proposal, but a treatment usually does not go into so much detail and is used to gauge the interest of publishers before committing to the more detailed document.

Unlockable content: This is hidden content (characters, objects or levels) within the game which can only be revealed by special combinations, achieving certain objectives or by the input of 'secret' codes released on web sites as marketing ploys. Unlockable content can add greatly to the replay value of a game.

Voice over (VO): Strictly speaking, a voice over refers to an off screen narrator who gives background information or other story relevant details. However, there are many within the industry who refer to all recorded dialogue as voice over lines.

Writer-designer: If the designer of a game also writes the story and/or dialogue, he's generally known as a writer-designer.

Zero-sum: Refers to a game like chess in which the general outcome is one where there is both a winner and a loser.

Gameography

The following is a list of the games I have worked on in various capacities during my game development career. They are presented here in the order I worked on them, though a couple have yet to be released at the time of writing.

Beneath a Steel Sky – Revolution Software, published by VIE.

Broken Sword – The Shadow of the Templars – Revolution Software, published by VIE/SCEE.

Broken Sword – The Smoking Mirror – Revolution Software, published by VIE/SCEE.

In Cold Blood – Revolution Software, published by Sony/Ubisoft/Dreamcatcher.

Gold and Glory: The Road to El Dorado – Revolution Software, published by Ubisoft.

Broken Sword – The Shadow of the Templars (GBA) – Revolution Software, published by BAM.

Broken Sword – The Sleeping Dragon – Revolution Software, published by THQ/The Adventure Company.

Wanted: A Wild Western Adventure – Revistronic, published by The Adventure Company.

Project Delta – Playlogic Games Factory.

Call of Cthulhu: Destiny's End – Headfirst Productions.

The Three Musketeers – Legendo Entertainment, published by Legendo Entertainment.

Agatha Christie – And Then There Were None – Awe Games, published by The Adventure Company.

Juniper Crescent – The Sapphire Claw – Juniper Games.

Mr. Smoozles Goes Nutso – Juniper Games.

Further details can be found on my website: www.incesight.co.uk

Index